DECLINE

AND

DESTRUCTION

OF THE

ORION EMPIRE

URIEL PRINCE OF THE REALM

(1981)

Unarius Past-Lives Therapy Students

(left to right)

Margery Vasquez; Brian Warfield; Michael Wilson; William Worley; Stephen Yancoskie

Unarius Past-Lives Therapy Students

(left to right)

Dorothy Ellerman; Thelma Fletcher; Roberto Gaetan; Franklin Garlock: Crystal Hampton; Decie Hook; Gordon Hook; Kathryn Hunt; Gordon Jenkins.

<u>Unarius Past-Lives Therapy Students</u>

(left to right)

Patricia Joiner; Calvin Kennedy; David Keymas;
Leila Landess; Dennis McNabb; Helen Moore;
David Osborne; Arthur Reed; David Reynolds.

Unarius Past-Lives Therapy Students

(left to right)

Ronald Robbins; Jennifer Ruth; Robert Sandlin;
Peter Sasak; Daniel Smith; Vaughn Spaegel;
Anne Swanson; Jeff Swanson; Hugh Taunton

THE DECLINE AND DESTRUCTION
OF
THE ORION EMPIRE

By

Ruth Norman & Unarius Students

Vol. 2

First Edition

Published By
Unarius Educational Foundation
P. O. Box 1042
El Cajon, California 92022

ISBN 0 - 932642-54-3

By

Unarius Educational Foundation

El Cajon, California

URIEL
PRINCE OF THE REALM
(1981)

Unarius Past-Lives Therapy Students

(left to right)

Margery Vasquez; Brian Warfield; Michael Wilson;
William Worley; Stephen Yancoskie

Unarius Past-Lives Therapy Students

(left to right)

Dorothy Ellerman; Thelma Fletcher; Roberto Gaetan; Franklin Garlock: Crystal Hampton; Decie Hook; Gordon Hook; Kathryn Hunt; Gordon Jenkins.

Unarius Past-Lives Therapy Students

(left to right)

Patricia Joiner; Calvin Kennedy; David Keymas; Leila Landess; Dennis McNabb; Helen Moore; David Osborne; Arthur Reed; David Reynolds.

Unarius Past-Lives Therapy Students

(left to right)

Ronald Robbins; Jennifer Ruth; Robert Sandlin;
Peter Sasak; Daniel Smith; Vaughn Spaegel;
Anne Swanson; Jeff Swanson; Hugh Taunton

Foreword

Although this biography - sad, woeful and horrible as it may appear to many - it does have, like all good stories, a positive and happy ending. For as you well know, we do now have on Earth, Uriel, the true, Spiritual Self of Dalos, who was never harmed, again returned with us as leader of the Unarius Mission. She is now dispersing by the hundreds, yes, by the thousands, these same negative forces that never die but simply regenerate their destructive acts, deeds and hatreds to others.

And when we speak of 'dispersing of these negative, astral forces', we do not mean in any sense of the word that they are being killed in their astral states but rather, simply that the fight continues, the battle rages even greater now in the Fall of 1979 as the Armageddon Cycle comes in full swing to be recognized, dealt with, objectified and these negative oscillations from this age-old past to be changed into positive oscillations with each individual who so partook of these acts, deeds and expressions of that long-ago past. So when we say, 'dispensing with these negative or astral forces', it is simply with the negative portent and content; more correctly: with the negative phase of the wave forms so created by these individuals within their psychic anatomies, to help free these souls to become, as they so greatly desire, freed of these pasts and to become a free-thinking soul - freed to choose their direction and expression.

These horrendous and drastic measures that were used throughout this lengthy, one-thousand year time toward this One Dalos, did create within each one, negative energies: That which he rendered unto Dalos. It has been proven countless times (as you will read in this text) by these students themselves: That which one does to another will and must return to him regenerated or, in this case, degenerated (more negative). It is exactly as our dear Brother Jesus told 2,000 years ago, *"The bread ye cast upon the waters must return to you multifold"*. Whether this energy is of a positive or negative nature, Principal prevails.

And so the astral world itself is being lessened of these negative hordes in their negative actions, for as they are being helped and healed by the One who was Dalos, (now Uriel), this does, in effect, help each and every one of the students so connected to these astral forces due to their own destructive pasts. Remember always, that energy does not lose its force or portent due to space or time but continues to oscillate in the self-same direction (positive or negative), in which it was first set in motion or directed.

This is what Dalos was endeavoring to teach these materialistic, obstinate people of the Orion Nebulae - mainly the Emperor and all those laboring beneath his rule: That which man expresses in any way, shape or form will, must and always does, return to him at some future time because the second or opposite end of the wave form is so firmly connected up to him just as he so created it.

So now this is the occupation, the mission, the duty and joy in which Uriel (the existing leader of Unarius on the Earth plane) is so involved and presently engaged: a twenty-four hour, round the clock mission to help these poor souls who for so many, many years did extend themselves in every conceivable manner, shape and form of torture to attempt to break down this soul and change his beliefs, for they had understood he was on a spiritual mission and desirous of helping to change their physical, materialistic beliefs and values in life, to raise their consciousness to that of a higher level which resulted in his capture and extensive torture (within the force field) during these many years. The many persons involved did so conjure up the many torturous means and ways to try to turn the mind of Dalos into their way of thinking but which was never accomplished, as you will read, for that is what these several books of Dalos are all about.

Important to realize for the reader is that as these students for the most part have so voiced these books to the class, it is that in their awareness and conceiving and becoming attuned to their own very destructive pasts were they so able to help cancel out these negative wave forms that remain within their own energy anatomies; thus healing was and is now being instituted as they become more aware of this negative past in a conscious way.

We can say it is a tremendous success story and that these Principles and truths hold true for anyone so tuning in to these testimonials, for the people as a whole on the Earth world have also taken a part in these atrocities in one manner, shape or form toward one or another of We Brothers of the Light. This is exactly the purpose of this Earth world: It is actually serving as a dump heap for these worlds who did not or would not get along in a positive way and who have refused time and time again to be taught by We of the higher Spiritual Worlds.

So once again we make manifest and available this life-changing, interdimensional science that will, for all who will study and conceive, change the lives of these persons for the better, for this high frequency so brought in at this time is carried in these dissertations through the subchannel, as the base plane frequency is of Uriel and the Brotherhood. Thus, these Dalos books can be a lifesaver to any soul so entering into the study of them, for as he conceives of the various acts and deeds expressed, (so voluntarily related by these individual students) and as he too senses that 'something', sometimes a slight or even greater emotion or reaction and yet again it can be a deep, gut-level nausea, he can well know deep within himself that he, too, did have such a part and that this act or deed has been and still is an oscillating energy within his psychic body. Through this recognition and acceptance, cancellation does occur and he can experience healing, just as have these many students.

So, dear student, do read with open mind, knowing that this is the 'open sesame' to your more peace-filled and better-adjusted future. We shall continue to serve you, the reader, with these bombardments of frequencies, these healing energies and radiations, just as we have unto this present group of students extended the power from the Minds on the most high Spiritual Worlds, and through the great Lens, re-generate this power which shall continue to come to you to aid in your great needs. This energy and Light from the innermost worlds and minds is surrounded with the Infinite Love of the complete Unarius Brotherhood and I, Uriel.

(Note: Should any reader (student) care to verify any of these relatings by the students by writing to them personally, any one would be most happy to verify his own testimonials. They can be reached simply by writing through Unarius.)

Chapter 1

URIEL SPEAKS ON THE UNHOLY SIX

5/27/79

Uriel: Good evening, class."

Students: "Good evening, Uriel."

Uriel: "Dear ones, this is a very, very serious time for everyone. I don't think any of you - not any one person - is realizing what it is that you are confronted with, for this cycle has been compounded many times. Again, the other day brought in the Amon Ra cycle when I was Ioshanna, the Peacock Princess, and most of you weren't receptive to the fact that it was another opportunity to work out. One other thing: The cycle of what we call the Orion Wars is now in focus. Remember that this went on for thousands of years, so this is not something that you can take a quick glimpse at and be all through with. Everyone had many parts in this and reincarnated back many times. On top of it, even after you left the physical and went into the astral, you still carried on this warring of the worlds in a psychic way, helping those warring, so it has been regenerated so many times that this is largely of what your psychics are created. And this is what we have to recognize to work it out.

"There is a great load here I am carrying with me because so many of you are actually fighting the Forces of Light unbeknownst to yourself. How many of you have been aware that things have been very negative lately? (There was a show of hands.) Well, of course, certainly! But what you do is to go along with the tide, you go along for a ride with this old negation. You don't change the phase; you don't stop it and do something to get out of that negation. There isn't an overabundance of energies just now, because it is gone to where I have to use my own physical, supporting energies to support my own psychic, so there isn't just an abundance now. We'll call on more shortly. But this is what we must all do to realize that this is our big opportunity: which is to realize that this is really the beginning of the "Battle of Armageddon". Everybody must really realize that this is our thousand-year battle, and we must realize that we are in it as a part and participle of this great battle! We must realize that we are confronted with negative energies, negative powers and astral obsessions from our own past, and only by recognizing and changing it, being aware (frequently being aware), can you help change this past, otherwise these forces will overwhelm you! You will be taken in by the great tides of negation.

"Now we do have great help for you but you cannot expect us to do it all for you; it just doesn't work that way; you don't gain anything by it. How many know what I am talking about? (Many raised their hands.) I wanted to come to the Center tonight to kind of shake you up a bit, to help you to realize the vast importances.

"Dear ones, let us realize that you all have been struggling, striving, living and reliving, reincarnating time after time, many thousands of times to arrive at this point. This is a crucial point for all of you; a very critical time in your evolution. There never will be a more critical point, and the Brothers really have been pushing me to tell of this. We must bring it to your consciousness so that you will become aware and wake up and try to walk more aware and more attuned rather than just going along in the groove, so to speak, of your past. I'm sure there will never be another cycle as great and horrendous as this negative cycle of what we call the "War of the Worlds" - the Orion Cycles. There can't be, there just cannot be! What could be worse than the destruction of entire planets? (Not only one planet but many, many planets!) What could be worse with murders of millions and millions of people?

"Yes, you all have made a certain amount of progress, but compared to the progress that can be made, it is trivia because the worst is now. The greatest, the heaviest cycle, the most negative cycle is now. So, when you people find yourselves, that you are feeling an opposition to me, realize what it really is. It is not me that you should be opposing, but rather yourself, because We Brothers are 'for' you, regardless of what you do, say or act. We are for you, not for your actions, but for you as a soul. We are trying to lift you out of this horrible debris in which you have been enmeshed. And frankly, that you have come as far as you have, is nothing less than a fantastic miracle for each and every one! You are all a living miracle - that you have come this far.

"Since the voice is no longer functional for the present, I am going to simply be aware with the Brothers, attune to them and let the powers flow. We know they are all in contact with us and they will help all possible, but these energies can't get through if you are in opposition; if you are blocked off; if you are resentful of any one of Us. So we will simply be receptive. We will be quiet and receptive to these powers that I know the Brothers are oscillating through unto us. Brothers, let the powers flow; let the greatest of energy be directed through this, our magnificent Lens to aid all possible, these bound souls; souls bound in their past. So, we receive now these healing powers and Light. Let the powers flow. A huge, blue ball of energy fills the entire room, and it encompasses the entire building. Now it must expand to include the entire city and the state, and of course, now, the entire world. Receive now these blessings of Light, this power of Love; receive . . . accept. And so it is. And so I trust that when I close this meeting, you will not revert again to the common way that you have been living and thinking and acting. Do try a little bit harder please, each one!

"Let us not have any intermission. I don't think intermissions are helpful. I think they dissipate whatever energy we do build up. I think an intermission for the two or three hours that we do come here is unnecessary. I think we should tend to our wants before we come. You see, we build the consciousness and the powers and the energies, and then to break it off in the middle, with an intermission, by the time you get into it again, people's thoughts are scattered.

"Is there anyone who was not receptive to the powers? Is there anyone who did not sense the power that was brought in? You all did! I'm glad, that's good. So we'll expect everyone to be a little better. But it is, dear ones, a very serious time. We cannot take it lightly. Anyone want to talk about anything in particular? What is it, Jeff?"

Jeff: "I would like to give a testimonial."

Uriel: "Would you step up here and do it before the microphone?"

Jeff: "The other day I came into the Center to put away the video equipment from the night before when we were filming, and there were two or three chaps sitting there. I had a terrible, instantaneous reaction to them. As I was cleaning up the equipment, I heard them talking in a way that I knew they were associated with a metaphysical group that is very predominant in the world today that is generally classified as the 'I Am-ers'. As I kept on working, I had to sort out what these feelings were because they were so very intense. I suddenly realized that I had been running into people like that quite a bit lately and they were representatives hanging on to the religion of the Unholy Six (Orion) times. This religion I believe was very closely associated with the computer at that time. The basic tenet indicated when they say 'I am that I am', is that they associate themselves with some tremendous astral force that they call God. Of course, it is not God at all and it is actually a statement of selfishness and separateness from everybody else. They actually believe they can channel these powers as I, too, once believed. I was quite involved in this when I first came to Unarius, and I know that those obsessions that were trailing around me gave Uriel quite a pretty bad time when I was here in California a few years back, although I didn't know it then.

"As I looked deeper into this, I realized that the religion of that day was the computer. What happened was that you tuned in to the computer mentally, through sequences of mental images, numbers, colors, whatever you wanted, and then you were given access to a certain position of information within the computer, not only information, but you were given access to buildings, basically to the society, through knowing these combinations of images. I was one of the priests or social scientists and part of my job was to have societies as modular units where, in broad terms, the program was laid out and then things were to fit together in whole sections. Then there were people who were below me who would work on the detailed work, and on and on and on.

"This is such a tremendously important cycle for me. I know this is why I am helping with the video at this time - because it is reversing that."

(Tom comes in, holding the heavy VTR and has nowhere to put it down.)

Uriel: "Where were you boys? Where were all these boys' minds just now? You see, Reyholds, and whoever is supposed to be helping with the audio-video, how blocked off you are? You would have had that stand ready for him if you were aware, wouldn't you?"

Jeff: "Do you want me to stop the testimonial?"

Uriel: "Go ahead, excuse us Jeff. Finish."

Jeff: "Helping in the video department is my way of reversing the programming that I put into this computer for the programming of the people thereby. The last few weeks I have been obsessed with watching television; I just can not get away from it or from the television movies. I'm always watching things and I rationalize it by saying, 'I just want to see where Earth people are at.' And that's pretty much what I did. I would sit at a panel and scan what was happening with these programs by inserting very broad-patterned questions into the program. I could see what the patterns of society were, simply by looking at how the computer was being used at any certain time. I read a book several years ago called "The Foundation Trilogy" by Isaac Asminov who is one of the ardent fighters against psychic phenomena. And the culmination of the book was that a galactic society was being molded and shaped by a society of priests who had this huge formula that they kept adapting as new conditions arise in society. I thought that was terrific. I could see all of infinity reduced to some kind of equation. I realize now that the equation I was so enthralled with were the equations I used to try to understand and manipulate the society of that time.

"I am in a position where I feel as though I have been Adolf Hitler, Eichmann and Hermann Goring and all the rest of them. And after thousands of years trying, I have been given an opportunity to repair the damage I have done. This is just the most fantastic time to get rid of this load of garbage that I have been carrying around for centuries and centuries! Right now I am going to be doing it by helping with the video, and then I will move on, to the next cycle. That is what I have realized."

Uriel: "Thank you, Jeff, very good. This is what we need to do: We need to become aware when we have these glimpses or visions or reactions, get in and analyze from where they come. Patty, you had one?"

Patricia: "I just wanted to tell of the incident when Uriel was projecting to us. I have been, for this last week or so, in this great, heavy, black fog of negation, and when she was projecting, . . . I haven't taken this seriously.

Uriel: "Honey, use your voice. You are talking like a little baby. Use your voice, you have a voice."

Patricia: (speaking up some) . . . "But I have to take this seriously. And then I felt this force just push me up, and all this negation was being pushed away. I thought, 'I must keep my mind on the Brothers. I can't let my mind go back like it always does; it goes back!"

Uriel: "That's fine, that's well and good, but don't forget it ten minutes from now. Come along, Roberto. Give yours quickly before the camera is ready."

Roberto: "Before I came into Unarius, I was very obsessed with mathematics and formulas and trying to find relationships. I really can relate to what Jeff was just talking about. My big obsession was with predicting planets."

Uriel: "With what?"

Roberto: "Discovering planets by mathematics. I went to a library and packed my bags - I had my room full of books and would take the information of the places of the planets and I would say, 'They are so far away' (indicating a distance measurement) and this is such and such, and I would try to find a relationship to see if I could come up with a formula in numbers. Mercury is #1, Venus is #2, Earth #3 in this numerical system and if these work, I could plug in any number and find out the position of those planets at any time I wanted because I knew the relationship system. In that way I could find the #10 planet, etc., etc.

"So one day I rigged up what I called a chart which takes me back to the Unholy Six thing where we could predict like in astrology. This is what I tuned in to. I took it out of my folder three days ago and I really had a good tune-in as to how I did work with these mathematics at that time. What I did was to come up with a relationship - this was four years ago - that worked. I did get a relationship and it was in the form of a sine wave with a line in the middle. That's what the relationship looked like. Then I said, 'Well, if I can get this right I can go over to another solar system; then I will find another solar system.' I would just plug myself all over the galaxy, finding relationships here and there. I began to become quite a maniac, finding just what I was into, in finding relationships. 'How can I predict the size of this planet, having never seen it before? But I'll bet you that it will be right because I have the formula.' I have always been very obsessed with putting the Infinite down into mathematical formulae, like Jeff just said. I am going to have to start working with this a lot more than I have. I just caught myself working this morning on another one of these mathematics, and I just let myself get taken over, instead of picking up the book and reading about Unarius Principles."

Uriel: "That's right. Mathematics is only the means to an end. It is not the end; it is a symbology. Before I leave here, I hope that some of these scientists realize this:Their mathematics, their algebra and all their high calculus and all their electronic media such as they use it, is only a means to an end. It is not the end; it is not the way. They have to replace all that with spiritual awareness; with the higher spiritual understanding.

"I'm really glad, Roberto. Roberto has really been in deep personal analysis for about two weeks. About every other day he writes me four or five sheets of all that he has been aware of and as a result, he is like a different person now. And this is what it takes - actual realization of what you did in one of the lives back there. And when you do change these wave forms, which he does, what a change! I have seen him the day after he completed the last one and he looked like he had just awakened from a long winter's nap. And this is what it really means to work out something. But much of the time, I must say, you people just go along for the ride; you simply relive it. You don't work it out. But I know that you are going to try harder now, in the future, aren't you?"

Students: "Yes!"

Uriel: "You, yourselves, can change things with a little awareness. So it was all very, very worthwhile. Well, you have much more time for more testimonials, and I am glad I came." (Film is shown after which testimonials continue.)

Kathryn: "I would like to tell of a little experience that I had today. I have just gotten really negative during the last couple of days and it has been building. I am totally unaware of anything. I guess you could call it negative insanity, to the point of telling one of my roommates that she had to move because I have been in such a power play. I have been wanting to dominate and control the people who live under my roof. I tell them the rules and regulations. It has really been horrible, but when I saw Margie this morning, I couldn't stand what I saw - a very blocked-off girl shaking her legs. This really disturbed me and I felt that she was in her room hiding away just like being in a little prison or a dungeon. I couldn't stand seeing this and I did let her know what I felt, that she was not welcome to live here. It has really been awful.

"But I went to the park and I was thinking and analyzing, when I came across this little robot. When I saw it, I thought, 'This was my past and what I am,' and I realize that I operate purely from my past and am just a little robot.' I thought, 'Okay, I'll tie it onto this little string that I have on my dress." I said, 'I'm going to wear this instead of my Unarius pin,' and I did. I put it on and I have really been blocked off for the past few hours. When Uriel sat over there, I was thinking about this little robot and I was really negative. I just got a horrible feeling with it. I don't know if I used to live with robots - I mean real robots, you know, mechanical robots or not. I'm not sure. It's like I am in a state of confusion, sort of. I don't really know what I am doing. I'm like a child playing with a toy. And this is the way my consciousness is right now. If I could get any help from the Brothers, I would appreciate it."

Dan: "Uriel has said, 'The only way out is in!' So I'm going in and I'll get out. While we were sitting here after Uriel talked, I really went in deeper into my consciousness. I became aware of some of the subtleties that I haven't really been aware of consciously, so I just want to polarize them. I realize these thoughts are coming from this Unholy Six (Orion Cycle) time and have been regenerated from life to life on down to this Isis Cycle, and even beyond that. When we were looking at the video, I had the thought that I really didn't want to see it. I didn't want to see people who were happy and making progress because I wasn't happy and I didn't want anyone else to be happy if I was not! I didn't want anyone to progress beyond me. I became aware of thoughts of how I have been looking at everybody (and not really aware of it until tonight) and looking at them as being beneath me and constantly judging them.

"This Isis Cycle represents my entire past with the Unholy Six up to the point that I have been able to conceive it. I've had a lot of realizations that I haven't talked about, because of one thing. I have been going around, blaming other people for what I have done. My guilt is so strong for blaming other people that I haven't been able to get up here and talk about it without repeating or regenerating these energies. But this basically is one of my worst problems. That is, I have placed myself under someone else. Who, it doesn't matter but it would always be someone else who could offer me a position of power and recognition. Then I have justified my actions of a negative nature by saying, 'Well, I'm only following orders; I'm only doing what this person wants me to do. Therefore, if it is wrong, it is his responsibility and not mine. This is totally false. This has been my escape - justifying the killing of Higher Beings such as Isis and Osiris and going around the world taking whatever I wanted or destroying whomever got in my way. This is the way I was in the Unholy Six times. I was a person who supposedly was a god and actually destroyed planets and entire populations and looked at them with the consciousness that these are things that are in my way - not looking at them as people but as objects. It was something like walking down the sidewalk and seeing an ant. If it was in the way, you would just step on it and not think about it.

"During the Isis and Osiris Cycle, Seth offered me a chance for a position and I took it and subconsciously blamed him if anything I did was wrong. This is a repeat of what I did during the Unholy Six when another person came to me and offered me a chance to play god. I took it to serve my own ends and then I cast all blame for my actions upon this person. In this way, I escaped the guilt for the actions that I took in these lifetimes, for awhile, blaming other people. I could go around and do as I wanted in my lower self, grab whatever I wanted, dominate people, suppress them or whatever I felt to do with them.

"And that's another thing: There are certain individuals here to whom I react more strongly than to others and those are the individuals who will, all of a sudden, rise up and snap back at someone if they don't like something they are seeing done. I will have the urge to go over and tell that person off for snapping back. I realized today that the reason is because I see myself in that person. I see that, first of all, I have dominated many people and kept them under my thumb when I was in a position to do so; I suppressed their desires. These people, themselves, became frustrated and resentful toward me and wanted to get back at me; they wanted to be able to express and I wouldn't allow it. So, I too took on the emotions of all these countless people whom I have suppressed. So it's double jeopardy. I'm getting back both this desire to dominate and this suppressed feeling that I created in the minds of these individuals. So primarily, the most important thing here is that I dominated others; then again in other lifetimes, I myself was dominated because I set the action in motion. I set the sine wave motion up by first dominating others, and then later, I received the reverse side of that wave form and myself became dominated. But primarily I react so strongly because I want to dominate. I want to be the one that's on top, and I don't want anyone else to have a say over my say.

"I have regenerated from life to life, a fear and a hatred for Uriel and the Brothers because at first, In the Unholy Six, I went along with the people over me and then I became dissatisfied. I wanted to be Number One and go out and be the only god, and so I ran off with a spaceship that had all this electronic equipment that could extinguish a world and the population upon it. I felt like I was the Infinite, but then it caught up with me. The people who were over me found me and I attacked them but I lost the battle. So I died hating these people who had suppressed me; the people who had put me in this position to be a god under them and dominate others as long as I would do as they, my superiors, wished. I, in turn, dominated and suppressed all the people beneath me. But then, when I tried to break away, I suffered for it. Well, I have regenerated this suffering, this fear of going against those over me.

"When I was Zan in the aborigine cycle and the Brothers came, I fell back into this reliving of memory of people over me who were so-called gods and had elevated me to the position of a god. I related to the Brothers in this way. I felt that they were gods and they had put me up in a position so when I went to my people in that tribe, I felt that I was some kind of god. My ego came in again just as it did in the Unholy Six times; the great difference was that these Beings really were Gods; these were the Brothers. But I was totally into my past, reliving these energies that I was somebody to be looked up to in feeling, 'I am like these Gods'. This is what I created in the Unholy Six, except in the Unholy Six group of planets, the people over me weren't gods either. They were just as negative as I was, but during the time of Zan, I substituted the Brothers in place of these negative, so-called gods. Then when I died because of my own egotistical actions, I blamed the Brothers for my death as well as the people who were killing me. However, the people who were killing me were, for the most part, people who were my own victims - people I had resisted and tried to dominate and suppress! So I died in that lifetime hating both the Brothers and all the people and many of them were you people who had something to do with my death. The irony is that it was all my fault. I brought it all down upon my own head.

"So again, in the cycle of Isis, I find myself doing the same thing. The Brothers, through Uriel (through Isis), tried to help me and tried to point out that I was reliving. This cycle came in again and I am reliving the energies of thinking I am some god and I am better than everybody else and that I should be listened to over Uriel, 'I know more than Isis!' In being like this, I cut myself off. I left the temple to serve another master and that master was my own lower self but I projected it out as another individual! I blamed him for any misdeeds so I could keep the sense of superiority and do whatever I wanted to do against Infinity. So whenever Uriel comes around, I fall into this fear, this subconscious guilt and fear. And why is that? It is because I erected a phoney barrier around my egotistical self, believing that I am some kind of higher being; a higher god and I am perfect. So when she comes around the facade is shown to me; it is reflected back to me. I hate her and resent her because I have to see myself. Instead of looking at myself and accepting my responsibility, I do the same thing over and over that I have done in other lifetimes. I point the finger at the other guy over me. In the case of the Unholy Six, there were those who told me to blow up a planet or said, 'You can be a god if you do this or that,' and also in this Isis Cycle, Seth said, 'If you follow me, you will have this and that.'

"Instead of pointing it back at myself and seeing my own phoniness, I have always projected it onto someone else so I could keep on playing the role of the Enlightened One; the one who knew more than anybody else, but I am the one who has suffered for it. I am the one who, up until right now, has been reliving these energies and thinking I am better than everybody else or that I know more than Uriel or I know better than anybody. It's all subconscious. Nobody would consciously desire to do these things, yet these energies are still there. It's time to become aware of them, recognize them and accept them because that's the only way I can ever change; that's the only way that the Brothers can break down the negative or add positive energies. In this way the negative experiences and realizations can be seen as a positive step toward recognizing the opposite experience - an experience of knowing I don't know anything. I'm just another small entity of consciousness with the Infinite that I am experiencing and these experiences are to be learned from so that when I have learned from this planet, I can go to another higher one. This will give meaning to its opposite; a meaning and knowing that all intelligence and wisdom comes from Higher Minds, and not from me.

"Gordon Hook said something the other day that helped me. He had a dream in which he saw himself reversing the polarities on electrical wires. Where the power was supposed to flow in a certain direction, he had reversed it to where it was opposing itself, sort of like the feedback principle of the audio system. This is what I have done. I have reversed the polarities of my consciousness within myself. I have, in my ego self, said that I am the Higher Self; therefore I am opposing the energies coming from this other self; this true self from the Brothers and I am putting myself out-of-phase. I am actually cancelling myself out with the Infinite; that is, until I become aware of what I am and what I have created and accepted."

Dennis: "This is a night to expose the phonys - me! I have been a student for three years and I think that in that time, I have learned very little of the principles because I have been too busy pretending that I know all about them. And this, through my evolution, has been my downfall. I have opposed the Brothers because I tried to pretend I was a Brother. I know this comes from the Unholy Six. I went to many planets and either pretended to be one of the Higher Beings or pretended to be one of the people from those planets. I built up this facade where everything that I said was a lie. I lived lies; my entire life was lies, falsehoods and facades and this is what I have been living. It has blocked me off from studying, because if I really studied, I would become aware that I was living without using principle. So it has blocked me off from really studying and that caused a vicious cycle because then all the fears and insecurities come in.

"Within the inner reaches of myself, I know I have this great lack of knowledge; I really don't know anything of the principles, yet that very fear motivates all my actions so I feel I have to prove how much I do know whether it is expounding principles or working with the video in any way, shape or form. My whole motivation is from these fears. This whole facade and all the levels of this negative bias that I am, also bring about the hatred for our Teacher because She can see right through that and She has pointed it out to me time and again.

"What I mean to say is that the hatred that I put out towards our Teacher is because she can see right through the facade. She points it out to me, as she has in countless other lifetimes including the Isis cycle when I was a priest in the Temple and plotted in every way I could to bring about her downfall and demise because I felt myself losing a grip on that power I thought I had. So I was instrumental in bringing about her death and the death of Osiris. But it goes way back into the Unholy Six and many cycles that have not come in that I have just begun to realize. But, that basic problem is of letting my thoughts and reactions just sort of slip by in consciousness, then putting a damper on it by saying that I have a handle on them; I know the principles. This is where I start slipping backwards."

Vaughn: "You have not even gone forward so how can you be slipping backwards?"

Dennis: "Well, by not going forward I am slipping backwards. Well, this is just beginning to uncover the many facades. I haven't even really moved in any direction. I am now just trying to strip off the veneer of my lower self."

Michael: "I have come to the realization in the past week or so that I am still an earth person. Although I am here, I still oscillate with a lot of the energies, thought patterns, etc., that I have always used. Looking at this from that aspect, it is amazing. In this group of people, during these testimonial sessions, I learn just as much by sitting and listening as I do studying, and probably more because it is a tangible reality. I just wish that these classes could be attended by many more people over the 50.

"A realization that should have been obvious to me as I didn't even look at the implications of that which I had been reliving even before I started working, has to do with a position I held during the time of our existence in the Orion galaxy. Many of the jobs I have had have had to do with the food industry, the third largest financial industry in this nation. But my best friend's parent owned a really nice restaurant when I lived in Iowa so I worked there; I worked at several restaurants and now I have another restaurant job. When I found out about this cycle, the first reaction I had was when Vaughn was speaking about people being used as food. I reacted strongly to it and said that it was dumb, people being used as food! If they had really high-level technology they should be able to make food synthetically but that is the way it was. It was really brought into my mind that I had a significant part. Today at work I was cooking (I am being taught to be a Jack-in-the Box), and this guy was coming on shift and he was standing to the right as I was standing over the grill. He didn't have anything to do and this gal told him to just stand there and look good. 'Yeah, but I feel useless,' he said, 'and I don't want to get fired.' The thought came into my mind to just throw him into the meat grinder! And I know that by laughing at it, I am still not facing the scope of it at all and what it involves, but that thought just popped into my head. It just showed me this valid past cannibalism we lived.

"The second thing has to do with this Osiris - Isis cycle. I have always had an affinity with Isis, the name Isis. I thought she was the Goddess of the Nile. I used to read about that particular person and I was always really interested. The other night I had a really difficult time getting to sleep for her name kept going through my mind. I had a very tumultuous feeling in me and I don't have a handle on my situation as far as Isis is concerned. But I had a dream that night and in the dream it was very strange. Most of us in this room were in it, and we were sitting at an L-shaped table and were getting ready to eat. We were dressed fairly nicely, and I recall a lady sitting three or four seats from me wearing a beige, tight twenties-type dress with a really long ostrich feather. I also recall Thomas (who had a rose on his shirt) saying something to me about his shirt being significant. I recall running into Vaughn who had a suit on. The last thing I remember in the dream is somebody saying, 'React to the Rabbi.' I said, 'There are no Rabbis in Egypt.' But the analysis that I have so far on this is that I associate Vaughn with a Jew, subconsciously and consciously too. Have you, Vaughn, ever been Jewish?"

Vaughn: "Shalom!"

Michael: "So, what I think this is trying to tell me in hearing, 'React to the Rabbi,' was that Vaughn was this high priest Seth and I was one of his sub-priests who went right along with the plotting and the killing of Isis and Osiris. I just thought I would polarize that. Thank you."

Ronald: "I saw where I had a lot of people who were highly mesmerized and zombied so that every movement I made, they did likewise. I saw also where I was involved in the television production which claimed to be from the Celestial worlds. Of course, it was really a farce. Today I had a very interesting experience. I was completely destroyed by spaceships. I had to even get rid of my clothes, so contaminated were they, and the chemical dust in the air and on the ground was just terrible. This was a very lonely time for me and a real test for survival.

"Walking on the planet's surface, I spotted a large group of children who had also survived the holocaust. They all came running to me and stuck closely to each other as the harsh environment demanded it. Eventually a rescue ship came and picked us up. The ego was such that I felt I got away with something because I survived. Now this may seem like a very negative experience to go through at the time, but I gained much through it and it is nice with the Science of Unarius that I can now reflect back on this. Previous to this experience I had no love for children. Being involved in the food processing at that time in one of the automated factories, children were just a source of food, to be ground up and put through the mill. I haven't developed a personal relationship with children from that life and subsequent lives thereafter. I do have a strong frequency rapport with them of strong guilt."

Crystal: "The pressure not to give this testimonial is so great I can feel the pressure of all those obsessions associated with this time and experience. Tonight, after Uriel gave the projection and we were watching the program, I had a tune-back into the past with the Emperor. When Vaughn dressed as the Emperor, the other day, I had a very interesting reaction. I was actually strongly attracted to that negative frequency! As I was sitting here I was thinking and for the past couple of days, 'How can I benefit in one of our plays by reenacting a person that I have been in the past; to really expose those obsessions and get them out into focus?' Tonight it came how to do it. My realization was that I was one of the many mistresses of the Emperor at one time and he had many. It was a great honor to be affiliated with the Emperor in this way. We would travel to the many planets. I would attend the blowing up of planets, like on a celebration. I see that in my mind!

"I had a dream last time we were going through this cycle, where it was presented to me that I was flying, up in the sky, twirling, doing ballet and singing and there was music. I was with Vaughn and others, but mainly it was Vaughn I was aware of who was there with me. At the same time, all of this entertainment type of consciousness was going on. There was an inner, very sadistic, devious feeling for the people beneath whom I was looking down upon. So the implications of this are very deep because this experience has regenerated down through my evolution.

"I gave a testimonial about three months ago, during the Hatshepsut cycle of a dream where it was relayed to me that this very serious throat condition I had was from using my voice on a mass communication system that was a prostitution ring. I thought this was from Lemuria but as the Unholy Six, we expanded our boundaries, not just from a worldwide situation, but all those destructive developments that took place on a planet encompassed a solar system like a cancer! So instead of having one part of a country used for something, it would be an entire planet that was used for some one purpose. Vaughn once mentioned to me in passing, 'Well, you know Crystal, there are whole planets that are like this. The souls there are strictly for lust and pleasure.' I associated this with the astral dimension but I am sure there was a certain part of the planet that was for sex and for prostitution, maintained especially for so-called 'love'.

"Now an interesting correlation has come up when I used to play this game where I was going to call for the weather of the planet to ask to have a sunny day or a rainy day, or whatever we needed. I called this planet the Planet of Love. Now I don't understand all the correlation here, but for our skit we are going to have a flashback segment where the people are going to reenact our time during the Unholy Six war where we were involved in weather warfare. So tomorrow when we have our meeting, I am going to tell an idea I have as to how I can reenact some of this past that I have been aware of and talking about, and incorporate into that facet of the reliving the flashback. But as I was sitting there, my whole body was trembling and quivering because of the strong opposition I felt not to get up and talk about what I did."

Gordon Hook: "Well, I did have a dream the other day that I related to Dan. It has to do with my stubbornness to accept things! I have been seeing this for quite some time but I didn't want to accept it. It was like I was just trying to place myself in the position of this situation. So in this dream, since I have been working on consoles so much (and this could have been part of my really big downfall), it was related to me in the dream that Uriel was in the middle of a console and I was on the outside and she was having me do a task for her. This task was very simple. All that was involved was a little tiny lightbulb and two wires. She was saying something about 'matchless'. This word 'matchless' stuck in my mind. When I was holding the wires - I don't know what I was supposed to be trying to do - but I knew if the polarity was changed in one of those wires so that the force was coming from both directions, I had fear that if those two wires were ever joined, they would annihilate that situation; the whole thing - the wires and everything around them.

"In the dream state, I touched the metal part of the light bulb with the plastic and it started to bubble and distort itself. I heard this voice in the background say, 'Look what you are doing!' I was sadistically laughing about it; I was happy because I was distorting this thing! So I awoke and went downstairs to write what exactly had happened and that's when it really started to hit me that this light bulb was the symbology of the flame, because the flames of our higher self or of our psychic anatomies are flames that are matchless; it is not one that is lighted by any physical means. I was in this physical consciousness so strongly that this is what I was doing: With the polarity pattern of the vortex and with this positive energy flowing from the vortex into the nucleus, I tried to feel that the positive energy was really flowing from me in the other direction; that I was the Higher Self and not the physical being. I became quite egocentric and egotistical because it was all self. When I changed the polarity of myself (I did this to myself), this is what I was trying to do: I was trying to annihilate my own psychic anatomy because it was getting to a point where, if I continued this way, I would go into the absorption cycle, what science calls matter and antimatter or their concept of it meeting and cancelling itself out, eliminating everything. This is what I was doing with myself.

"Also, during the video tonight, I felt like I was having one releasement after another. One of the things that I remember was when they were showing the crystalline cities with the beautiful lights on them. I was seeing this through the Lens, and the sparkling regeneration and the frequencies coming from it. I had to admit to myself that I had been blocking off seeing this psychically. I didn't want to accept it because, if I accept seeing it and bringing it into this physical consciousness, then I have to accept that it is real. And I have been so physical and fighting so much that I have just about totally destroyed myself, up until this point. But now I want to change it all around. I feel much better inside.

"There was another thing related by one of the students about radiation on this planet. They entered into a belt of atomic radiation. The planet had been under atomic warfare and there was all this radiation unbeknownst to the ship that penetrated it. But as I watched the "Space 1999" program the other night, they had entered what they call a dust particle belt, trying to get to a planet. In watching, I attuned to this radiation because all of a sudden I started to itch all over. There was really a very bad itching and my first thought was of radiation. When I thought about it for a bit, all the itching left; it ceased completely! I have not in the past related things that I do see (whether positive or negative), but from now on I am going to dedicate myself to relating what I see and to not hold it in.

"When Uriel was up here and she had her Dove of Peace gown on and gave a projection to us, before she even opened her arms and showed us the gown and was standing there with her arms folded, all of a sudden I saw her arms open up showing every feather there was. The fingers pointed outwards and there was a golden shaft of radiant energy encompassing every one of us. It was just as real and just as solid to me as everyone of you sitting here right now. Thank you."

Jeff: "While the filming of Severus was going on, I was really deeply in tune with Isis and Osiris and got into my usual little thing of hating Stephen. I was ready to kill him, I really was! I was so angry that I didn't want to be around the video equipment; I was afraid that I was going to destroy something. I finally went home. When I got home I realized, 'Well, Jeff, here it is! Isis and Osiris and you just left again! You are not doing your job,' which is what happened in the time of Isis and Osiris. I shirked my duties and was very self-indulgent. Earlier in the day, I had nearly gotten into fisticuffs with him over the standard for the banner I was carrying around. I had been working hard on it for a couple of weeks and it was just about finished. The last touch was to put this flying saucer on the top and I was having a hard time with it. I spent a couple of hours trying to find the right glue to bond the saucer to the top of the staff. I finally did it and it was about set. I walked into the back room with it and he walked up and yanked it off saying, 'That doesn't look good; we want feathers there.' He just about got it right there but I realized, after a good deal of introspection that this Isis and Osiris cycle was a reliving of the Unholy Six time where I commanded a fleet. You see, I developed an ego about this standard. It was my standard, was the reason I was so involved with it. It was I, expanding the frontiers of the Orion Empire with this fleet! Now I believe that during this time of the Unholy Six I was doing too well. I was becoming a force to contend with, so I was forcibly retired. They didn't want to eliminate me because it would discourage others from doing good things for the Empire, but I was retired and went to one of the pleasure planets that Crystal was talking about where I completely indulged myself. But my ego drive was so totally one hundred percent involved with my standard and my position, my job, and everything like that, that it was the same thing as a death warrant to me! I was shattered. It was interesting that this thing of 'ripping off the standard', happened in front of the Emperor. He was standing right there!

"So during the time of Isis and Osiris, when Stephen and I incarnated into the same family, the talk of spaceships came about. I began to become very, very insecure about my ego structure, who I was, my sense of self-importance and very carefully gauging any other (Stephen), 'Is he higher in estimation in our parent's eyes?' I was always judging and keeping score all the time. At the same time, I became very irresponsible and began to draw on these energies of this 'retirement plan' (I will call it that for lack of a better name though it was a bordello plan). But that was the energy and the source of most of the hatred I have had for Stephen. I'm sure I spent the rest of that lifetime, dedicating every waking thought to trying to eliminate Stephen! Everything and anything I could do to get him, was what my life was all about from that point on, having a little fun, but of course, it didn't do anybody any harm but me and I'm very grateful to be relieved of that burden that I loaded myself up with. I am beginning to see Stephen now as a brother of Light."

Crystal: "I had another realization that there were women who were abducted. When planets were taken over, some of the women were brought to this pleasure planet; they would be trained and forced into this type of sexual activity through drugs and programming. I remember during the Atlantis sessions, Uriel made a comment that she watched a movie where she had seen women go into this supposed place where they were to have their hair done, facials, etc. They would sit beneath the hair dryer, and unbeknownst to them, they would be programmed into their subconscious, 'I am a pleasure machine. I am a pleasure machine.' So I am sure that much programming was done on this planet for women. I have had this thought but I would say, 'No, it is just your ego. What are you trying to do here?' But I can remember being one of the leaders of a planet that did this type of thing but I didn't understand it at the time. All I could get was that I was a woman who was one of the leaders of a planet that used men as slaves and the women as sex objects and to stimulate sex impulses. I know these things are true because I didn't have any idea of this until I sat here now. It just dropped into my consciousness tonight and solidified."

Patricia: "I know what Crystal is saying about this planet is true, about the brainwashing because I have a very sickening feeling in my stomach. And my basic problem has been that it has taken all my energy to get my mind off hair. I try to attune to the Brothers and some hairstyle will come in. What this represents to me is these people coming in to get their hair done and we brainwashed them for this pleasure thing.

"The testimonial I have relates to my inability to communicate with people. I say, 'Now I have to get out with the students more; I have to do more things and not enclose myself,' but I think, 'What do I do? What do I say to them? I can't make conversation.' The reason I can't hold a conversation is because I have caused so much dissension in trying to separate students. In the Isis and Osiris Cycle, I would try to separate and cause dissent between people so they would break away from the Light, just to destroy the Light or to try to. With this realization, maybe I can work better with the Valneza planet group because I have been causing dissension there. Any idea that has come up, I have immediately opposed it; in my mind I haven't liked anyone's ideas. I haven't even liked my ideas! I'm going to make this effort, like I said last week, to communicate with the students and not to look at folks in such a negative way. So what, if I can't communicate? I will learn."

Gordon Jenkins: "This was one I had a few weeks ago but I didn't want to relate it, 'Wait until I get something else.' One day when Uriel was sitting in her chair with a few other students nearby, talking to us, she asked me to get her a drink of water. So I filled the blue glass with water - the glass I had bought for her. So I had a big thing within me about drinking water and her. When I gave the water to her, she said, 'Oh, this water tastes so good; it's pleasant and crystal clear. You people must have gotten all the gook out of the fountain.' I was very shocked when she said that the water tasted good because it has never tasted good to me; in fact, I could not drink water from our fountain it hasn't tasted good to me. I could not drink water that didn't have some purifying substance in it!

"A bit later she asked for more water and as she was talking, Decie started to reach over for the glass but my hand shot out like a bullet and got it before Decie could reach it.

Uriel then said for me not to touch the top of the glass, so because of my tremendous guilt I got out my shirt tail and started wiping it off, when she said, 'Oh, no, not that! That's ten times worse!' It didn't hit me right then, but later, in thinking of my reliving, the situation occurred during the time of Dalos when we rarely gave her water and when we did, it was always dirty and served in a filthy glass. That was the time when she was imprisoned by us. I felt bad about that."

Gordon mentioned seeing a student with her black and blue face lying on a cot in the sewing room and described his great state of shock at the sight.

Vaughn: "It is very simple, Gordon. You were now seeing your victims and that is why you were so emotional."

Gordon: "I felt that either I or my men had beat her up in her cell as she was lying there. That was the relationship with me.

"When Ronald talked about food processing, it reminded me of something. My first job was in a bakery. In that bakery, when we had left-over donuts we always put them in boxes near the oven to dry out so they got hard, after which we would put them in this gigantic grinder for crumbs. Then we would put it in with the good dough and sell the donuts with the crumbs and people never knew what they were eating. This was like "Soylent Green", passing the human food that people didn't know they were eating."

Jennifer: "This Unholy Six Cycle has really helped me take a good look at the energies I have and the facade I put on of insecurity and which I am, but in a very quiet and unassuming and very sweet way! The key word for me in this cycle is 'control'. During the time of Dalos, I was involved in many ways in programming because I have been very involved in typing. I have a fear of electronics which I didn't want to face. Although I feel an affiliation with it, I have actually blocked it out of my psychic. But with the typing, I actually feel this sense of power with the fastness of punching the keys, and I actually have a little set-up in my house where I sit next to the window and it overlooks the students who walk back and forth. It is actually like being in a control room! People wave up there, and I am busy typing away.

"Most of all, yesterday we had a group meeting for our Valneza filming and I was so blocked off, I really didn't like any of the ideas. I felt everything was unintegrated afterwards. Everyone was into working out this weather karma but I felt I didn't even understand this at all. I went home and thought about it. I started to see, first of all, I was jealous because of Crystal being the polarity and I was trying psychically to put down any idea. I would come up with these little ideas, but the energies I was using were really to control everyone in the group. So I was telling another student what the ideas were for Valneza, and as I was talking and stated what we were going to do, it all sounded very integrated. I started to realize, 'Boy, where was I? Just sitting in the back trying to control everyone psychically.' And that is why I have a lot of problems when I react. When I criticize or judge and start to feel negative, it's usually because I'm trying to control the situation. I'm putting this limit on myself which I did to many thousands of people. Even in my job, I stand over the people as a waitress and regulate when they are going to have something to eat, or whatever.

But they are in that position of sitting down and taking just about anything I have to hand out. So I have really got to take hold and turn around these negative energies of trying to control things which is only controlling me at this point.

"Another realization I had was stated in the Unholy Six pamphlet, about the planet for breeding. In this cycle, in the last week and a half, I have had an obsession; it would be an insane thing but all of a sudden I will go and eat something very fattening, then I would bloat up at night. The night of Isis-Osiris, I felt pregnant; in fact, Dotty asked me if I was pregnant. I said, 'No!' But I actually have a great fear of becoming pregnant. I started to look back because this is a pattern that has happened in this lifetime; it will just come in and I won't do anything about it. I realized that it started when I was baby-sitting in high school. I would refuse many of the activities in high school so I could baby-sit! When I baby-sat, I would usually eat. I had the whole neighborhood selected with who had the best food, but it would be the guilt with these children. In my mind, one of the worst things is giving birth; to me it is close to a painful death, and that's really the way I looked at it. So it shows the guilt I have had with all these ladies or people in my psychic and these babies. So, that has really hit home with me.

"Also, reading in the pamphlet, I've come to the realization of how badly blocked I am and have been, and how deeply I have buried my soul. This is the greatest chance I have to start to put things together. In this lifetime, I wanted nothing to do with science and I thought that the nine planets in this solar system were just about all that existed. When

I came into Unarius, the last thing I wanted to accept were the 32 planets. Most of the students take right up with the reading, but not me. So, I am very grateful for this opportunity to start going back and seeing this destruction that I caused these souls. Thank you."

Stanley: "First I would like to thank the Brothers for the reading I got last week about the radiation poisoning. I feel about 200% better mentally and I am just about over the physical distress. Last night Jeannie called and told me about a television movie that was called 'Five Million Years To Earth', about an ancient Martian invasion. These Martians were giant grasshoppers. I thought, 'Yuck, I have done that in the past.' Today I started thinking about it and remembered an old television serial called "The Invaders". One of the shows was about a way they were trying to invade Earth through these herds of locusts. With millions and millions of locusts, they were going to invade Earth and completely wipe out the crops. I remember another movie I saw about the cockroaches that invaded Earth. Last night in the lab, I turned on the light and I swear there was a cockroach the size of a mouse; it tuned me in. I know now there have been lifetimes where we have bred bugs, locusts and such. Jeff also gave a testimonial on this. We would invade the planet and destroy the food so that the planet would have to ask the Unholy Six for assistance because they would be totally helpless against this invasion.

"Another thing: After Uriel's talk, I was thinking how I really react to women who have their hair up really high, like they have in a restaurant with the big bonnets on their heads. I really react to that; I just don't like it.

"Another thing came to mind, that Jeannie and I have been more or less recruited to do the program on planet Dollium about this pair of twins. These twins, whenever they are in contact with Uriel are always wearing big caps or bonnets that I react to. And I remember the first realization I had about Atlantis; I was a scientist in the positive part who helped to bring in a teaching device that fits on the head. In the negative part of Atlantis, I used this teaching device to put wrong or negative teachings into the minds of the people. In the Unholy Six, I believe that I, myself, wore this cap. It could be used as a teaching device because it was directly hooked up to the computer but I used it as a genius device. I could put any information that I wanted in and say that I knew everything. I have had a similar problem where I thought I knew everything. That is just about a general problem I think - the big ego. This is the reason: Because I had this cap, I thought I would dupe the public by making them think I was all-wise. Then I got to the point where I actually thought I knew everything by wearing this cap. I thought just now, as it comes to mind, that I was actually in tune with the God Forces! Just like on the planet Dollium, that is the reason these two beings wear these caps; it's through frequency relationship that these energies can come through.

"Until just recently, I have had a very large problem in getting up in front of people. I am going to dedicate myself to not letting these obsessions stop me from expressing myself and in polarizing these things. I have had many realizations that I haven't told and for all I know, they could help somebody else.

It helped me a lot. Sitting here a few minutes ago, I could feel this force in my solar plexus just pounding, 'Don't get up. Don't get up!' 'No, man! I am going to get up and I am not going to let it stop me anymore.' So, I would like to thank the Brothers very much."

David Keymas: "I have a pounding plexus and a big complex too. I haven't realized the implications of this realization, but it has been blocking me off from certain endeavors I have been undertaking. At times I look at this light here and my eyes will become prismatic. Many prisms will seemingly radiate out. As I squint, my eyes can control these beams of light, these rays. I have had this thing, and would pretend that I was projecting light to people or zapping them with a laser gun. I would squint my eye and focus this light to create prisms and cause the largest prism to radiate down into a person's head - this all with my mind. It is not really happening, it is an optical illusion as far as physical reality, but as a mental reality, it is real.

"I borrowed the laser and have taken it home while working on some filming. I have been trying to do some little special effects in making these Higher Beings appear. I was really having a good time doing this, trying to make them appear in a burst of laser light. I couldn't quite hit them right because I had to hit the beam right on this little angel I had drawn. I realized the reason I couldn't do it, in a sense, is because I was using negative energy. I didn't really accept the fact that I had fired laser weaponry at the Brothers who had incarnated during the climactic part of the war. There were many Brothers destroyed incarnating, just like in the Renaissance Period, to help avert the great negation. In this Unholy Six cycle, as it stated in "The Epic", Uriel and Michiel incarnated down and were destroyed many times by the negative forces, of which I was one. Hopefully, this is going to wake me up a little to what I am doing; I am not going to get anywhere until I realize what I am doing.

"Mal Var came over and borrowed the laser because he also wanted to use these special effects with it. When you do the special effects, the laser actually reflects back out and will hit you in the eye and I get a headache every time I do it because I am actually looking into the laser light. One is not supposed to do that. Mal Var got it right in the eye, and I felt very guilty. I said, 'My gosh, I should have warned him.' He really suffered quite a bit from this because it really hurt him. I feel that in some lifetime, of which I am actually getting the full realization right now, when I was under him, and he was very adept at working on some project, I did away with him in some way. This is how I feel because he has been helping me in various ways and there is a subconscious resentment because he is so adept at what he does and I haven't even begun to begin. Hopefully this will wake me up to what I am doing and to greater realizations. Thank you."

Steve Sandlin: "Last week I gave a testimonial wherein I said I had been the leader and the coordinator and a right up-front man for a situation that was happening during the Unholy Six. Roberto got up the week before or after that and said something to the effect of his desire to dominate. I realize that in doing this play for Severus, the part that I had in the play was listening to Vaughn while he was Emperor, I have sensed a tremendous desire to dominate and I found myself in this play, planning my part out so that I would be in a dominant character! I was actually enjoying the fact that I had a dominant role, but what I was doing was placating this desire to dominate and lead. I realized, while I was sitting back here listening to Vaughn talk, that I wanted the position that he was talking about, the praise and prestige that one would have when he came back from this mission. So, I realized that this desire had been my motivation most of my life.

"I also want to verify my own realization about an experience Jeff related tonight. I have had a couple of dreams and realizations that there was a religion and a philosophy built around this computer. We would change or alter our need for the particular planet we were wanting to control; whatever we wanted to do with it. In this dream I saw that I was on a planet, in a huge plaza or courtyard with many people. There was a building on the far wall. I didn't have any sense as to why I was there. I wasn't questioning what was happening. All of a sudden, everybody started racing for this building. I started racing with them and again it was without any wonder or question. There were people to the side of me and to the back of me who were freezing dead; stopped in their tracks. Everybody was racing toward the building and when I got to the building, all of a sudden I became aware of what was happening, that this was a regular routine done at a certain part of the day; something they didn't question and was part of the religion. It was green, fluorescent, a building and inside were the people all bowed down to this one wall that looked like a glorified circuit board of fluorescent green and that's when I started asking questions. 'Why do these people just accept what is happening?' I had the impression that at certain times of the day or week, this high frequency sound would come over the planet and the people would start racing toward this building. Those who made it would live for another day or until the next time this happened. The people who were frozen in their tracks were later picked up and taken someplace. I didn't really know where, during this dream state, but I can now see it being the meat or food factory.

"What brought in the realization was a mention by Uriel that the worst possible thing we could do is destroy planets, and I could really relate to that. For me, it is the worst possible thing to think about - this religion that we formed around this computer and the psychological techniques. If we destroy a planet, the persons can come back, but if we destroy and warp those people's minds and their concepts of themselves and of the Infinite, that's the worst possible. That's where my great karma comes in and the fact that not only were we destroying physical planets and physical bodies, but we were actually warping, distorting and twisting these people's minds, putting concepts in them via this computer and this religion we developed.

"I have this thing with the Lens, where you are supposed to tune in to it for help for your problems. I sit here and say I'm tuning in. 'Well, I'm ready for the answer. Now give it to me,' but nothing ever comes. It is like this Lens is a vague concept. I can see where this is my remembrance of this computer! We would teach the people to tune in to this Lens or the computer, 'That's the God, that's the Godforce, the Godhead.' They would tune in and we would program the information we wanted them to receive through this computer into their minds. In that way, through this religion and the computer, we were controlling all these many people, on all these different planets, according to what they needed to be. We would have the pleasure planet programmed to do that. Whatever we needed to do with that particular planet, we could do through these philosophies that we set up.

"I was setting myself up as one of these teachers or priests of this so-called religion. We were filming early this morning, dressed in our costumes for Severus. Edna was with me as we went over to Winchell's to get donuts, and as we walked in with this crystal on my head and this dress on, some guy comes up who has his two kids with him and is outraged with both of us. He says, 'Who are you? Are you Jesus and Mary Magdalene?' I couldn't understand what he was so upset for; it was like he was protecting his kids. I now see that was what he could conceive in his mind as Jesus and Mary Magdalene, but it was those energies from the past when we had set ourselves up as Saviors. As a matter of fact, he made the comment, 'I thought Jesus and Mary Magdalene came to this planet to save us.' I could see that we had set ourselves up, going to these planets with this religion and this philosophy and our computers to 'save' these people. In reality, we did just the opposite! Thank you."

Jeannie: "At the start of this Unholy Six Cycle, I had a dream which was a really good learning experience for me. I was sitting on a windowsill with a friend. It was as if we were in a skyscraper because as we looked out all we could see were clouds and the sky. It was as if we were waiting for something. Sure enough, there were these rainbow colors in the sky and I said, 'Oh, there are the Brothers!' I was looking excitedly, and she was going to jump off. Suddenly I saw this black strip in between all of these beautiful colors of the rainbow. I was horrified by it and realized that it was obsessions. I grabbed my friend and said, 'Oh, it's obsessions! It's obsessions! It's not the Brothers.' Then surely enough, that disappeared and these beautiful colors came down which really were the Brothers! I learned from this how obsessions are directed by the negative forces to come in all forms and incredible disguises to fool us. With this Unholy Six Cycle, I know that I am just skimming the surface of this negative past that I have lived. I am beginning to see the real mud that is in there that must be gotten rid of. I am going to fight. I'm going to get through it. I've got all the help; it's all here. I just thank Uriel and the Brotherhood with everything I can give for this help and the beautiful opportunity that we now have to be able to get out of this hell world that we live in!"

Leila: "I found out that I am quite a big phony too. I go around with this mild demeanor and everything but it's really a big cover-up. Because of my great negativity from my past lives, I don't want to face it. In order not to face it, I act like a mummy. I just don't let go of my emotions; I can't. At home I seemed to be able to let go of my emotions when I was raising children, but they were not the right emotions. My whole life has been hell. I know now why it has been. I haven't been able to face that. I have blamed it on other things. I have been blaming you students because you have been negative toward me. Well, now I know that you are not to blame for that; it's I who am to blame.

"Last Monday night when we were having rehearsal, Thomas, out of his love for me, was trying to get me to wake up to what a mummy I really am, and to my lack of expression. I accepted that because I knew it was love that he was projecting to me. The next day I felt good about it, but then the following day, I became very negative about it. All the students seemed more negative toward me. I thought, 'I am just going to have to become a home-study student, I can't take this any longer. I can't do it.' Thursday I came up here and I was going to leave a note for Vaughn that I could not be in that procession that night. Danny was here, and he was sitting at the desk in Vaughn's office. I told him I was going to leave a note for Vaughn that I couldn't come tonight. He started talking with me and I felt his love too. He let me know that I had to really put out a strong effort to be there and not to let my lower self keep me from coming. I said, 'Well, I will come.' And it was a very beautiful evening for me. Uriel looked different to me. For the first time I actually approached her, and I know that it is all within me, now. I have been blaming all of you. I have been like a whimpering dog at the master's feet. Because, when I went up to Uriel to talk the other night, when I got through crying and talking to her, I thought that I was just like a whimpering dog; it was sickening. I'm such a phony that I can hardly believe it myself. After listening to Uriel tonight, I realize that I am going to have to really straighten out and start using principle because this is the last chance I am going to have. If I have any intelligence at all, I will start using it."

Stephen: "I realized for the past few days I have been in a rather common consciousness. I realized that throughout my evolution, I have tried to be anything but common.

"Regarding the Isis and Osiris Cycle, Vaughn asked me to print a sign for him to use in the scene that was going to be filmed. So I immediately took it on to do. Interestingly enough, I said, 'Oh Vaughn, how are you?' because I felt really good about Vaughn, so I proceeded to make this sign for him but it was a very negative thing. The first thing I had to do was make two lions for it and I couldn't think what a lion looked like. I tried to draw one freehand and it looked like a mutant. It was the most ugly thing I have seen. Then I realized the negation and finished the sign. Anyway, when I was working on that, I got a note from Uriel to make a sign for Isis. I didn't think anything of it. I took it and started getting an idea for that, and Uriel said to me, 'Vaughn forgot all about that sign.' Here she had told him, two weeks before, to give me the message for this sign! All of a sudden it hit me: Here I am doing it again - being duped by this guy, seeing him as good old Seth. 'He's such a good guy. He is trying,' and my not realizing what he was up to. I felt really ridiculous when I realized that I was backing him in that lifetime. Then I had to make him up for this reliving, and again I was jazzed on this. I realized with this Orion sign, I was obviously promoting Vaughn back in this lifetime.

"And to go into how I realized more of that: I just picked up a book that was going around the Center here. I picked it up and started reading about this guy who happens to be fabulously wealthy and rides around in spaceships, and he is so bored with his life that he decides to go out and have some fun causing commotion in the galaxy. This is what I was, in one lifetime anyway. I had so much money and everything as far as materialistic needs (kind of the intergalactic playboy type of thing), and I came in contact with Vaughn.

I saw him as, 'Well, he's got some interesting ideas here.' What started out for me to be leisure-time activity turned into actually backing, in a financial way, this whole thing.

"When I made his face up, and you know how you are into something and you are into the technicalities of a job and you are just in there, knowing that you do this and do that - kind of marking out the facial lines, not paying attention, especially when you are doing something in a creative vein to the whole thing - then you step back and look at it to see what you have achieved. When I did Vaughn's makeup, doing his eyes and makeup and everything, I wasn't paying too much attention. But as I stepped back, my stomach just did a flip-flop. He put on that costume and I just could not believe. I never thought that he was under all that or that this was he! It was like, 'What did I get myself into?' and was the whole feeling. But again I was into it up to my ears; a name to hold up and all of that. In order to save face (the old save-face routine), I stayed with him, and stayed with it. Of course, the obsessions backed it up even further.

"One thing more is about what Jeff stated in his testimonial about the retirement thing that I relate to so well. I really felt this was part of my duty - to eliminate people who were maybe a little too over-zealous or ambitious. This was one thing that I had to do with Decie and Calvin and a couple of other people who have been in a lot of the video. I was told to relax them for a couple of the videos, but it wasn't like a rub-out thing. Even Decie said to me, 'Gee, I feel like I am being sent out to a retirement farm.' So, I really do relate to that, and I relate to it in such a way, that it was such a very impersonal thing. I have been feeling lately, a checklist of consciousness, 'Check this off. I have got to get rid of these two to the retirement farm, etc.,' so this is a little bit of a start for me. There is so much that I did want to get that out. It's very important for me to realize that I was right there backing Vaughn. I know it's true because of the things I have been doing, but I don't yet know the complete implications of this horror."

Peter: "In this life of mine here, I have a habit of trying to tell people how to live their lives. In other words, I size them up and say, 'If he knew what I know and could understand what I am going to tell him, he could change his life. Instead of being unhappy, miserable and sick, he could improve his life.' So I have practiced this in my life many times and have used my will on my children. I have tried to pressure them into following my ideas, the people that have worked under me also, and that has been my pattern through this life. Before, I didn't recognize but upon coming into Unarius and studying, I have begun to recognize what I have been doing. I said to myself, 'Look, what right have you got to tell people how to live? It's their life, it's their responsibility. You're not a teacher, so try to let them go and keep away from it.' Even with that knowledge I have had the habit of falling back into that because I look at different ones here in the group. I say, 'Well, so and so ought to do this, and so and so ought to do that.' What I am bringing up is that Vaughn says that if you want to get into this Orion Cycle and this working out, just think of what you are doing and concentrate on something that you are doing and what you have.

"So, in reading the dissertation on Orion, I came to the part where the people of Orion were going out to the other planets and taking them over. Some of them they destroyed. Others they took over by sending in people to work their way into the places of influence, and eventually changed the conditions of the people. The thought struck me right then and there, 'There you are. You can see your own pattern there. You were one of those who went out to the different planets and were put in there to infiltrate. You did your part in changing the way of life of these people and made their lives conform to the dictates of the Emperor and of the Orion philosophy!" (Uriel: "Good awareness!")

Brian: "I just realized when I sat down, the obvious thing that I hadn't realized, was having to do with Vaughn. All of a sudden there was an immediate obsession in my throat when I sat down. I thought, 'Well, something I said wasn't right!' Then I knew instantly that I really believe something I said had to do with my reaction to Vaughn and the fact that there was one thought that there were all those forces behind me and that I could kill him. At the same time, think how ridiculous that is! I'm trying to say that I am as negative as you are. Well, I've got as many negative forces, almost. I've been almost that negative, but I realized that really it was a fear of knowing that he could kill me that I reacted to so strongly. It was a feeling of, 'He can!' What came to mind is that he did because I wouldn't have had that reaction. It's the same thing with me as the over-zealous person who may be showing through a little bit that I don't have any respect for the guy that I am following. I'm trying to get my own power. I believe in one of these many lives that is what took place. It's something that is good to come up with. It has been the first time in the last many weeks that I have recognition of certain reactions in myself that I have had for Vaughn which I have always just powdered down before. But real resentment and hatred came about, and it is good for me to at least recognize it and face that fact - that they're there. I sure won't ever get rid of them until I do. So that's where a big part of it is, no doubt. With an ego that big, he wiped out my chance to fulfill that ego drive, which would be an incredible negative reaction."

Vaughn: "So here's the culprit. Yes, this is all true. I admit to every facet of every experience that everyone has had. The basic truth of it all is, that no one would have any relationship with anyone else unless they wanted to and walked into it. But that's not what I really came up here to talk about.

"I have had a very disastrous, yet actually a very wonderful awakening. And the awakening has finally reached the point where I can feel the damage, the great destruction, which I have caused in many untold numbers of people. When Stephen made me up for Tyrantus, (I just realized that is a similar name for the name of the polarity on planet Basis), but it's a very interesting feeling and experience as I was being made up. Stephen was doing it but he was not doing it from his conscious mind. I don't think that he indicated that. He was doing that under guidance, because this whole reenactment was done under the guidance of the Brothers. I opened up to it, and I indicated that I ought to write in a part where I ought to indicate my responsibility in the 32 planet reenactments, and that somehow, or in some way, it must be indicated that the Orion people and the Emperor were responsible in the ultimate. Even though I wasn't in my conscious, so-called lower self at the time, it came about bit by bit through the assistance of various students, for instance, and they were all influenced by the Brothers . . . Gordon Jenkins who came up with this picture or photograph of Emperor Ming, the Merciless, from the old Flash Gordon series back in the 1930's; Decie Hook who indicated that there was a good picture for Tyrantus; Jeff and Ann came up with the idea of the long fingernails. Up until that time, I was trying to determine what the appearance would be. Then after the makeup that Stephen put on, I didn't even bother looking in the mirror. Stephen laughed, 'That will be enough out of you; you have been discharged already.' I didn't want to look in the mirror because inside of me I knew that the actual person would be there that I was attempting to face and objectify. That was the first basic reason. I looked in the mirror and saw that, and didn't say anything; I immediately felt that it was I. So, there was not a disguise, Stephen; it was really not a disguise but I was actually tuned in to the actual personality of this individual. I was completely inphase! It was almost as if he stepped into my physical body. I had no difficulty being that person. I even enjoyed being that person for that time. However, after it was over I hurriedly got rid of it. I almost tore that costume off; I pulled the bald-headed wig off; I pulled the beard off and the mustache and I wiped everything off. I just did not want to be in that consciousness because I knew how easy it would be to stay in it.

"Then this is the big change; the tremendous awakening but again, it is only a beginning. The awakening is that I was forced to face who this individual was. Uriel was Ioshanna at the time and again set this up as she did everything else and had me confront her and admit to the past crimes of Tyrantus, and

Satan, and all the other people whose lives I lived. You know very well that when you are standing up here talking and giving a testimonial, when you are just speaking words, you know deep inside you there is something you can't get out and that is, you can't relate it because it would practically do you in. At least this is my feeling in all the testimonials I have given.

"Well, last night I saw a film that was full of bloodshed. Uriel turns these films off immediately because they are a very poor excuse for a consciousness of any high kind, but she purposely left it on for me. I am the one who got the benefit from it. I mentioned in chit-chat that something happened. What happened was that people were killing each other; it was the kind of situation where a ship was being taken over by some pirates. The one person would kill another and the pirate would be killed, and the good guys would kill the bad guys and the opposite situation. I laughed at a particular burst of machine gun fire. I actually laughed even though it was just like a titter, when Uriel turned around and looked at me. She said, 'You think that's funny?' I thought about it and it was as if somebody slammed me firmly right against the wall. This is the way I felt - flattened with the realization. I have had it before but this time the realization of the shock and the trauma was such that I actually enjoyed seeing these people being killed! I was watching it as if it was fun. That was the first time I could really feel what this insane person has been in all of these thousands and thousands of years, all of these various people who have headed armies: Napoleon, Hitler, and all of the Army Generals of the United States. In spite of the fact that they say they work for peace, they can't help but have some sort of sadistic feeling. And there's no feeling whatsoever; it's part of the game and that's the way I felt.

"Tonight, all of this has a great meaning when the projection was made for us, and all of those who accompany us. I saw this tremendous beam of energy shoot out of the Lens, like a trap door opened and a bolt came out into our psychics and to the rest of the world as a whole. I saw three flashes like that and right afterwards, I saw how many lifetimes on the earth world have been of one great, mad obsession. There once was a film called "The Magnificent Obsession". That's what I saw. In my mind's eye I saw the earth worlds as one whole state. There must have been hundreds and hundreds of lifetimes that passed by, and as it was one big continuous play that was constantly repeating itself, I saw how, just above that (all of these plays were acted in the dark you might say, acted in half-light) there was some light flowing, and above that were the true worlds; the real worlds that were in complete spiritual Light. The feeling I had was that I saw, in my mind's eye, this grand obsession of living in this physical world. I was beginning to see the outside of it, to recognize that it was an obsession; it wasn't real life; and that I have been trapped through my own actions in this world of illusion.

"The first true and sincere feeling that I have had in all of this destruction was this feeling of actual repentance, and I could actually feel inside of myself, some tears! I have been a very hard-hearted person because that's the only way you can get to the top - crush the souls of other people. You can only project this onto other people as a whole.

"Gordon Jenkins has always stood up as an example of a person who has been the executioner, the grand, great, high executioner. Well, Gordon Jenkins doesn't stand a chance, in any way, although he has his responsibility, for acting out whatever job he has and jobs he was given.

"So this is the whole fact here. It is very hard to talk in a concise way because I'm not trying to give a speech. We are all tied together - your hates and your resentments and all of these emotions that you have toward me, toward anyone else, but particularly it is I - are very evident and very credible in that I have, as Stephen pointed out, given the appearance of helping you, where in essence, you never knew my true identity. When Stephen pointed out how he wanted to help my grand design in some way, he didn't realize what was hiding beneath the veneer of that individual. After the power was attained, the beast emerged and he overwhelmed everyone. That's the history of my evolution. This is a crucial time in this Unholy Six Cycle when we are coming to the beginning of the causal factors of our diseased minds, at least in a physical way. I feel almost as if I am going to blow apart with realizations; I couldn't contain myself, blowing apart with the awful realization and all of the true realizations that there has been a constant fight. Even though Uriel and the Brothers are here to save our souls, I have been fighting, even her help! There are times when this is not true, but generally speaking, the fight has been against accepting the truth of the identity of being the worst personality in terms of what we call a human being. To have to fight and face and to constantly strip away one facade from another, this stripping away has been fought with tooth and nail, so that identity would not be uncovered, yet that uncovering was that much an attainment towards living in the real world. That speaks the truth of how deep the layers have been.

"It has to be the truth: that if one has buried himself so deeply within the unrealities of an insane world, he must have been totally insane, totally obsessed. So here we are to face ourselves. Thank the Brothers (I was going to say, 'Thank God'), thank the God Forces, that we listen to our inner self! I am really truly apologetic towards the Brothers of the Infinite, the Unariun Brothers. I have not been able to say this before because my guilt and ego were so strong, having been a Brother, then absconding, and then turning on my own Brother - the worst crime one could commit. I don't know if one can, but it has been millions of years of struggle to come to the point where I can actually turn around and face myself as I did on Thursday by actually walking into the consciousness of that person I had been. That is the time when the power that was held by the Emperor was the greatest because he controlled hundreds of planets. That was a very important time to have had that confrontation, and only the wise, Intelligent, Spiritual Beings would know this and this is why it was set up.

"It wasn't done in conscious mind, but it proved something to me. I had actually asked; I had opened up and said, 'I have got to face myself.' And they helped. I took one step and they brought it all together. So, friends and associates and enemies of my past, let us not hold onto our dislikes and emotions. Recognize that every time you feel angry at someone, it is the past that you are looking at and that is the first evidence of recognition: the minute you feel emotional. But the answer is that we have got to study; really crack the book. Open it up every morning and night, and during the day. The first thing when we awaken, we should, all of us, spend the first hour in study, and the last one! All I can say is, 'Thanks!' "

(A Note From Uriel: "It is doubtful that there is or has ever been one so thoroughly set against We of the Light as has been Vaughn. He used such force, effort and great length of time to prove his righteousness in his egomaniacal ways, being the very leader of the evil forces. For one such as he to be able to view his past, to accept these countless, destructive deeds for eons of such living, through these eight years time, is truly a most incredible proving of principle of the Infinite powers projected into him. Vaughn is indeed now on his more progressive way and no one is more joyful for him than is myself. He has been the most trying of students but we do see success; a progressiveness that must ever be present. There can be no turning back now, for if so, he is lost forever . . . to the very demons of hell. So watch it, Vaughn, constantly!"

* * *

3/26/80

A Word From Uriel

Although mention has been made regarding Stephen Yancoskie - how after sitting in with the class of students for two years, students who were all working out and reliving their drastically negative pasts - although his frequency seemed to be more positive than most, he would nonetheless add to the others' testimonials as is part of the class sessions. Stephen did, from time to time, add his testimonials of destructive or negative deeds and acts which he had expressed on the Orion planets, even as to going so far as to speak or teach against myself, (Dalos), even though she (Shimkus), went purposely to the planet to aid Dalos in his Truth-bringing and Light-bearing.

It has taken several months for the entire picture and truth to be known. Now, it is shown from the Brothers on the inner, how the Orionites very cleverly, and through the sympathy and compassion of Shimkus, and with the use of their advanced electronics and mind controls, actually got Shimkus under their control and influence to the point where She was working with them and against Dalos and the Light! Once one came under such dominion, it was impossible to extricate himself. So Shimkus thusly was used in this overpowering way to aid their cause - yes, against the Lighted One, Dalos, who was striving so constantly to show the leaders of the Orion planets the error of their ways and how it was (and is) vitally important for man to begin to recognize his supreme and rightful heritage - his true Spiritual Self - and about which, these people knew nothing and lived purely the material way as does the lowly earthean.

So Dalos' plans were greatly thwarted, for now, instead of having his polarity and companion aiding his cause, she (Shimkus) was actually opposing it and Dalos, due to the electronic controls they instituted, and the sly, conniving ways they used to get Shimkus to their side. In their lies, they had her believe that they were understanding and going along with what she and Dalos were teaching, but they very cleverly and deceitfully turned the tables on her and she became their victim for not only many years, suffering much torture beforehand, but each lifetime Shimkus returned to meet with or be the forerunner for Dalos, she would again relive the negations and regressive acts of the past! The same Orion leaders would, in one way or another, trap, ensnare or trick Shimkus and hold her on their side, preventing her from joining her polarity.

"Arieson relived these deeds continuously for several months. He strives to recognize and become aware when those negatively influencing frequencies are present or oscillating to him, but it is and must be, a constant awareness with him; otherwise he is, in thus reliving, once again opposing or rejecting Dalos in the present!

But as in all cases, the past will and must reoscillate into the present and when one becomes aware of these principles, so can he little by little, work out this past and find the freedom he desires. Arieson is making good progress in this direction but realizes that there can be, lurking behind any bush, so to say, that dragon of the past - the remnants of the destructive acts of the leaders of the great Orion Empire, who treated my companion and polarity as a prisoner, using the worst kind of war tactics to cause her humiliation with use of their hypnotic efforts, to cause her to believe that Dalos did not want her or her help and many other such heart-rending acts.

One of the most drastic measures they used was to make Shimkus into a sea monster, actually cutting the jaw bone, etc., to build in the gills of a fish, and through their advanced medical skills, actually turned this lovely, young girl into a sea animal, similar to a fish, and successfully forcing her to remain in a huge water tank for many years in succession. Cruel doesn't nearly describe their actions, but now these persons are all reliving their parts in these atrocities, as is the principle of regeneration or degeneration. And to free oneself of such deeds is not an easy or quick process, and minus the Unarius principles, an impossible task!

So the wheels of the Cosmos continue to turn and those who are ready to come along to gain the understanding can change their lives for the better way. And we say, 'We do have the way!' The know-how is contained in our many volumes and to those who study with sincere effort to learn, we can and do extend that help, that healing balm of radiant energies that changes all unlike itself to Light.

No one person must need remain in the negative stance that he is presently reflecting. The Infinite Radiance and Light of the Inner is for all mankind, regardless of his present position in life. If we can help to change the direction of one such as Satan - which has been done - there is no one who is a hopeless case. But one must, himself, extend the effort.

But what an example this soul, Arieson, has been; that he has actually been cut off from his true inner or higher self ever since that first fateful day in Orion, over one million years ago! Yes, sometimes the karma is heavy and long-lasting, but what a lesson one learns! After all, this is what life on Earth is all about!

Needless to say, perhaps, as mention has been made in various places, but the Spiritual Being Dalos, who reincarnated twenty-six times into the Orion planets to try to help bring about the needed change, was also captured and tortured on many occasions. On one trip,

Dalos was entrapped within a force field the Orion scientists (so-called) had built; a small area surrounded with a force field, served to imprison Dalos. This capture existed for many years with added measures of various torture means, of just any possible, conceivable way, to cause discomfort and pain. All this, of course, was done to try to pry from Dalos (and Shimkus) certain spiritual powers which the Orion people believed was some secret unknown to them. These higher powers were, of course, the result of inner development and were not possible to pass on to another by any physical means.

However, these Orionites were impossible to convince and did dole out much constant torture to the two Spirit Beings before finally killing them outright. Of course, these acts were repeated life after life, and it is these same souls reincarnated on Earth. But a true 'aware' Spiritual Being would never 'give up' or renege, regardless of what means he encountered. And once again, Dalos (Uriel) is making an all-out effort to help these criminals of the past unlock the numerous destructive acts and deeds that have locked them into their own dungeons, many of whom are, even now, quite insane as a result.

The help and way out is again being offered and shown, and some certain progress is being made by many; yet with karma as heavy and long-lasting as has been the lot of this group, it can scarcely be expected to be eradicated in any brief time. Yes, Dalos was killed in various ways during his many thousands of years work on those planets. The present life on Planet Earth is but a repeat for Dalos, now named Ruth Norman, and it does seem as though we shall get through this teaching cycle without the physical torture and murder. So perhaps some progress has been made! However, man is still quite animalistic. It has been but a few years ago that President Kennedy and his Senator brother were shot by some radical. Man destroys what he does not understand or what he hates.

However, this 20th Century does bring to mankind the better, higher way of life for whomsoever will. Man can learn to become a spiritual being for it is the very goal, objective and purpose of life. Man must begin to forgive himself for his past errors by recognizing and admitting them; thus cancellation can occur.

- Uriel

* * *

TESTIMONIALS

3/25/79

Dear Uriel: I have read at least a dozen times the Unholy Six as revealed through Vaughn. Through much conceiving and analyzing as to my parts in connection to what was done to Dalos, several realizations came to me.

Those realizations caused me to feel that my main part was in the use of sound as a means of torturing Dalos. It was the mini-reading concerning my time in Atlantis as a scientist in the experimentation on human hearing that brought about the rest of my realizations. Due to my hearing impairment (nerve deafness), I can't hear sounds of a high pitch. So I feel this may be due to the use of ultra-high frequency sound for torture. One thing that really solidifies this awareness is my hearing test and using my voice to annoy others.

Due to these realizations on my part toward Dalos, it really hurts when I'm aware that Uriel was that same person. I can truthfully say right now that this relating of such horrible times in the worlds of Orion, especially my destructive lives, is the biggest bombshell. It really makes me wonder how Uriel and the Infinite could be so forgiving and let me exist.

While writing this I feel so terrible and tearful and it is so hard to express my feelings. I'm trying to remain objective and keep control of myself.

Sometimes I'm lax in my study and practice. However, I'm still trying not to give up. I must keep on going.

Sincerely,
Harold Wieland, Delaware.

* * *

3/30/79

Dear Uriel: After reading the dissertation on "Wars of the World", I feel that I had something to do with eliminating the people on the planets used for fuel and food. In this present lifetime, I have had trouble with hemorrhaging when I gave birth to my daughter and later my daughter hemorrhaged after a tonsil operation. I am on blood thinners now because of a heart operation. I feel I killed people by giving them large doses of blood thinners which caused them to bleed to death. I am now receiving the other side of the wave form.

I thank the Brothers for helping me to see my past which I have buried so deep. All my love,

Mildred Grunewald, Ca.

7/24/79

Dear Archangel of the Universe, Uriel: I've read and reread the latest Unarius publication, concerning the Unholy Six Star System of Orion and it's staggering and horrible, almost beyond comprehension for a mind to conceive. And, yes, I was one of the perpetrators!

I have always been sensitive to changes in temperature, especially these days, involving my right side. Therefore I may have been involved among the culprits who cooled the air molecules causing freezing sensations to Dalos, then rays of great heat would be brought in, causing great agitation to him, (Dalos who is Uriel this present lifetime).

Yes, I feel I was one of those of Orion who knew of all these plans and secrets, ways to torture the great ambassador, Dalos, a beautiful speciman of an earth person. We lied and spread distorted concepts that she was distorting truth and that the Orion culprits were the innocent people instead which was the reverse of truth.

I realize now that a great plan, a purpose in life must be completed, or static energy will be the result and I realize, since reading the latest Newsletter, that I set in motion a plan to help Unarius with funds which hasn't been completed causing me great frustration and guilt, hence all the depression, heart tremors, etc., that I've suffered.

In the rest homes (2), in which I spent some 5 months, I saw many people in varying states of decay of both mind and body. No doubt many or all were my former victims during that horrible Orion cycle.

Although I have been told I wasn't a

scientist in any civilization, I have been a very negative individual, willingly doing the bidding of others for gain. I may have advised and schemed to distort the truth, when I knew better.

This is my confession and I hope it will help the thousands of people I have previously harmed by leading them down the garden path. And I do ask the Brothers to help me as I struggle to keep my sanity in this insane world. I need inner strength to accept and dismiss my horrible past. I hereby re-dedicate myself to the Infinite Truth.

I'm sleeping much better than I have for the past 2 years or more. Love,

- Student Ruth Trager.

* * *

7/24/79

Dearest Uriel: I have been very busy reading and rereading the latest Newsletters I receive from the Center exposing the horrible parts in which the Unholy Six Orion people were involved so long ago. I feel so badly and I want to cry so desperately that it's almost more than I can bear because of the horrible realization I have had regarding the roles which I played negative parts in Orion.

I realize that I must involve myself in this working out by exposing the guilts (and karma) for which I am responsible, the horrible deeds I performed against humanity and Spirit during that Orion Cycle.

I was one who served in an administrative way under Emperor Tyrantus and was responsible for propagandizing the moves that would be developed politically to inform people of their purpose in relation to the Empire.

I realize that I can only point the finger of guilt at myself. I am guilty, too as far as Dalos is concerned, for I knew of the plan and purposes that all we guilty people were hatching out for a long period of time. I'm guilty, too, for helping to detour and prevent the plan of Dalos when he attempted to polarize the planets.

I now realize the strong tie with my son - that our tie will now be resolved - also with the conspirators and spies with whom I was involved during that cycle.

Love from Unariun,

Ruthie Trager

* * * *

7/26/79

Dearest Uriel: The other day after being at the Center, I developed a severe headache. I realized it was from the Dalos Cycle. Upon awakening the next morning, the realization came in that I envy you! It started when you were Dalos, when I monitored your every action and became aware that you were godlike. I envied and I resented you for this.

I also realized that I monitored every reflex action of eyelids, etc., that you displayed and that is what I am reliving when I watch practically every move that you make when you sit in front of us at the meetings. I actually get embarrassed for doing this.

I felt so good about these realizations and objectifications. I thought that this is what I have been waiting for so very, very long (many years).

Then tonight, after the meeting I closed my eyes when you touched me on the forehead and couldn't meet your eyes. I was so shocked by this action, when I walked out I thought, 'Where will I go from here? Am I controlled by the dark forces?'

7/28/80

I awoke this morning feeling confident that I can become victorious over this Battle of Armageddon. I have so much love and respect for the Brothers, I am not going to let all their efforts be in vain where I am concerned.

Uriel, I adored you from the first time we met, when you looked into my eyes and I teared. It was such a beautiful experience for me.

Therefore, I could not understand it when I became negative toward you when I became transfixed in my parts. It has been so painful and frustrating for me that I could hardly stay in Unarius.

To feel that this negative energy had been changed and then to become negative again is still a shock to me; however, I know that it will be worked out and I will be stronger for it.

I also intend to overcome the childlike expressions that I have and become the woman that I want to be.

Thank you for reading this.
With love,

- Leila.

* * *

3/15/79

The most important thing on my mind is concerning the Orion Cycle. Your letter thanking me for photos and the question you asked: "Could it be splattering of blood on hands and feet?" What an impact that made and then a great flood of tears!

This raced through my mind, "I really killed all those people; not only did I carve the bodies but I slashed their throats!" I was not ready to admit that I did that and I put the idea out of my mind. But when you asked that question, I went to pieces and had to do more objectifying and now have finally accepted and I just had to get this to you, for the battle of the old self is a hard one.

Well, I'm glad I got this off my conscience, and do feel better about this! I'm sorry, but the battle of self really is a tough one. I know you had everything to do with this objectifying, recognizing and finally admitting and above all, accepting this.

My thanks to you and the Brothers for all the help you are giving me. Oscillations of love to you,

- Nellie Gonas, Pa.

* * *

May 23, 1979

Dear Brothers and Sisters of Unarius: My husband and I are not compatible in the ways of our spiritual interpretations as he does not want to be bothered about any philosophy or even discuss it. I feel that we are reliving something or other towards each other at the present time. I have had feelings of a tenseness or some kind of void in our relationship. There have been times in my life, in recent years, where I have felt psychically attacked and it was a real struggle to pull out of it and several times I felt a force pushing me when I have fallen down and gotten physically hurt. It must be astral obsessions. I'm hoping to cleanse myself of them. It really takes effort to overcome depression and be positive more of the time.

My knee, hips and back joints are better now and I have been able to adjust the joints better, especially in the lower back that has been causing the trouble. I have read the information you sent, 17 days in succession, as much as possible and have felt healing energies from it, for which I am deeply grateful. Thank you so much.

Sincerely,

Avis Perisho, Ca.

* * *

Dear Uriel: Since I am back with Unarius now, I am feeling much better and no where near as miserable. I am sure that if I study hard I will recover from this slump I have fallen back into completely. I enjoyed very much reading the testimonials of the students and *Encounter with Life, Death and Immortality*. In the future, I will send for the remainder of the Unarius Library.

James Elli, Ore.

CHAPTER 2

May 30, 1979

INSPIRATION FROM THE BROTHERS

Subchannel: "We are going to talk about a subject that is very important to all of us but first I want to show this article Vaughn found in the Enquirer. It says, 'U.S. performs successful head transplants'. So, Like we say, if you don't like your body, or it gets old, just take your head off and put it on a new body. What is so funny about that? Why is that so funny?"

Stan: "The way you said it."

Tom: "How did I say that? How would you like me to say it? (In a serious tone he resumes:) There was a head on this body . . . gruesome! The reason I say this is that apparently that was what did happen in the past or there were experiments to try to prove that bodies could be exchanged by beings and that body could be dissociated very easily and still maintain the normality of the person's life. You no longer have that same oscillation because you've got two person's psychics now, no matter how hard you try to get out of it.

"So I won't read this but maybe Vaughn will let you look at it, individually, if you want. They have done this with monkeys, and they have done it for many years, in history-making experiments. Monkey's heads have been successfully transplanted from one body to another and kept alive up to seven days. Even more incredible, scientists say it's medically possible today to transplant a human head from one body to another, a miraculous breakthrough that could overcome disease and stretch man's life span. To what lengths they will go to stretch the life span of a man in this Earth world who is, more often than not, in total misery in his karmic way of life!"

Dan: "About two weeks ago I saw a movie on television about this man who had his head transplanted on another man's body!"

Tom: "Was it drama or a horror movie?"

Dan: "It was an hour-and-a-half movie."

Channel: "Well they are saying that it is science fact now, not just science fiction. So, it's very true.

"There is one thing that we might want to bring up to the students of Unarius about certain misconceptions of this concept that might be appropriate at this time to bring out so that we can gain a better understanding. When we go through particular pressures of our past, we are prone to blame this situation on exterior environment or other people; that we have been pressured into this by some circumstance or in some other way and means, we have been put in this position. Everyone is to blame except the self.

"We have said many times over, the Brothers have given us this one vital and pertinent principle that no matter where we are, no matter what we are doing or expressing, it is because we wish to be there. We have discussed this in many ways and through many avenues, and yet, turning around this concept again, we can say that the lower self or the ego consciousness cannot - we repeat - cannot take the blame for anything that it does, by virtue of the state of consciousness that had been built in by that person. Therefore, no matter what it does, the ego is perfect; that's how it sees itself - perfect! This is why the ego will not admit to any wrongs done by that self.

"So, to compensate for some misadjustments the person might have or might be confronted with by other people about certain things he is doing wrong, he might feel a twinge of guilt. So he must try to reproject that guilt out to another person by saying, 'You were in this particular situation with me!' or, 'You were in this lifetime and did such and such an act and were just as negative as I was!' or, 'I see you as a person who did this or that,' etc. Well, all of these are ways and means to justify, from the ego's point of view, that it cannot be wrong and even in the justification of the self, trying to explain away why it is in that situation at that particular moment.

"What we are trying to say is that it all boils down to this: The Unariun student in general has the wrong idea about how to deal with his fellow Unariun in the sense that we should never attempt to tell another student that he is, in the student's wild imagination, some barbarian, some executioner, some murderous person that he sees this other person as, or to somehow include that other person in his own reliving. One cannot do that and truly stay on that plane of self-introspection because it is the modus operandi of the ego to continually try to exert influence in the direction of others, so therefore he will focus the attention on others so the ego will not have to defend its perfect position. There is a lot of this going around and it's getting pretty much like a disease and is totally misunderstood by the students of Unarius en masse.

"Do not attempt to put another person in a place where you want him to be. We have heard this both firsthand and from other students. Uriel has many times said this same thing over and over, so it is nothing new that we are saying here tonight. Some students try to give a reading for other students and ninety-nine times out of a hundred, that student is only seeing his own past in another light! And because he can't face it, he is trying to place the other person in that particular role to get some kind of psychic revenge, or to again refocus the attention on another person's negation rather than his own. It is the ploy of the ego consciousness to continually try to empower some esteem within himself. He is continually trying to evaluate himself on that 'perfect' level, and any time that person sees his own actions in an imperfect way, he has to justify that. The ego demands that he justify it because, as he feels, the ego is perfect.

"So therefore, in justifying that the ego is perfect, it must constantly siphon off all of the imperfections in himself onto others, and there are many imperfections in each one of us. So when he does that, it is very easy for a person to say that you are a so-and-so type person. Most Unariuns will say, 'Yes, I am that type person,' easily, because most of us are. Most of us have these emotions. So in doing this, the person who is siphoning off his own emotions onto another person and qualifying him as having that particular feeling or as doing this thing in some past lifetime, he is in fact strengthening that ego consciousness. He is continually keeping that perfection intact by siphoning off and telling another person he is negative.

"So we must not try to 'read' for other people; we haven't even learned to 'read' for ourselves! We hear this many times, 'I saw this person as being with me as a negative commander in another lifetime.' Well, it's up to that other individual to understand that he was that commander, even if he was. It is not up to one to judge another person just because we have no objectivity. More often than not we try to put the person in that place, out of superior feelings, or desiring superior feelings, or desiring revenge, or out of desiring some kind of equalizer. Maybe you feel inferior around that person so you want to constantly put him in the position of being negative, from that past.

"So let us start getting to the point where we are only judging ourselves and our own actions. If you see another person in another lifetime as doing a certain thing to you, more often than not you are the one who did the deed to him. Our concern in working out our past is not only what happened to us, but what we have done to others because there is nothing to work out in being a victim compared to being the perpetrator. There is much to work out in being the perpetrator of some negative crime you might feel you have committed in some past. Being a victim, if we know energy and how it operates, is because we desired to be a victim because of our guilts. We have placed ourselves through our guilts, subconsciously speaking, in a position to be harmed.

"So how are we to forgive the harmer if we, ourselves, wanted subconsciously to be harmed to begin with? So there are many deviations of how the ego expresses itself. And we have to be very, very careful if we want to be truly introspective. It's a very dangerous game, this self-analysis, because we are so often fooled by our own attempt to re-justify the value of our worth to the Infinite. Someone came up to me the other day and said, 'I have got to do something to justify my existence.' Well, if he were inphase with the positive side of his nature, that train of thought would never have come into his mind. Do we honestly and truly feel forced into many things we do in Unarius? How many feel forced into things they do in Unarius, not by others but by their own nature? Forced? They are not doing it out of love. They feel their inadequacies, fears, frustrations and all of those things. (A big show of hands.) Well, I raise my hand further than the rest of you, and if everyone was truly introspective and honest with himself, he would say this is the case.

"At this present stage in our evolution, our very actions prove that we have expressed egotistically in our past and we have expressed noncreatively. This has been proven to us many times. Truly creative persons do not live on this Earth plane. This Earth plane is a plane of experience; it is a school, and many of the Eastern philosophies have looked at it that way for many hundreds and thousands of years; many mind science groups feel the same way about it. But the point is, we must start to analyze and use a little introspection before one says, 'I am a Unariun, therefore I am creative,' or 'I am a Unariun, therefore I am negative.' How many people honestly feel and truly feel they have very little worth and value and a very negative past and have thoughts of not being a good person? How many people truly feel down on themselves all the time? Honestly and truthfully, where is our consciousness? Because it is just as negative to us in our soulic evolution to be down on ourselves and beating ourselves all the time. This is not the purpose of this great, illumined Science - to continually bore a soul into the ground. It has not the purpose to make you a defeatist person or to make you a person who is so low on yourself that you do not know which way you're coming from.

"This Science is an uplifting one, a resurrecting and an illuminating Science; therefore, why all the frowning? Why all the frequent emotional outbursts by us all? Why do we feel ego-deflated every five minutes? Why do we feel that we don't have a grasp on our evolution most of the time? Why don't we have that peace of mind most of the time? We have really got to start a new cycle here and start to analyze these things because if we don't watch ourselves (since we have been so negative in the past), we are about to admit that our life isn't worth a hill of beans. That is equally a sickness, just as the highfaluting egotist on the other side of the pendulum that we swing to has a sickness. Do you understand what we are saying here? We are saying that you can detour your evolution just as greatly by beating yourself all the time as by thinking that you are the salt of the Earth. Either or both ways will detour you!

"Your self-evaluation should come from learning the Infinite Energy Principles, and that is: Every wave form, every constructive thought and form is a part and parcel of the Infinite Intelligence and there is no way we are going to get out of it, even though we beat ourselves and strap ourselves down to some mental flagellation. We continually try to tell ourselves that the horror of our psychic anatomies is so black and vile, that no being in his right mind would have any connection or any association with us. This is the way we treat ourselves. This is just as horrible to our evolution and just as much a deterrent to us as that other egocentric self.

"Now, I'm sure we oscillate between the two extremes. If we are not tremendously high on ourselves, we are tremendously low on ourselves. We have to learn to gain an equilibrium and perspective of our own value. All that vileness and all that negativity is part of the Infinite. Even the greatest egotist is part of the Infinite; whether or not we like it, all of that is part of the Infinite. The only reason we would not like it is because we have not learned the opposite side; we do not understand we are at the present time, unable to do anything except judge it, condemn it, revile it and vilify it, to create a world around it that must somehow justify its existence. And we call it hell, but it is actually our consciousness that we have built this vile energy around, and what we call our 'defense mechanisms'.

"The ego itself is a partial defense. So if it is perfect - and the ego is perfect, is it not - it must be perfect to exist. Anytime the ego learns and accepts that it is not perfect, it dies immediately. How many times have we fought some realization over and over, day after day, week after week? The pressure builds up and there are these continual jabs at this defense that we have built up - this ego. It is continually justifying its existence. How many times?

"Then when we immediately say, 'Hey, I wasn't right, there,' the force immediately goes away. That ego has been completely annihilated for that particular moment. Now you can build it up again like an inflatable tire; put some more air in it and you can inflate that ego full of hot air. There's nothing there, but you can constantly inflate it. But it gets deflated and there is no force there once you come to that realization that it's not perfect or it is not right as right and wrong are like black and white. You find you are not right, so you have got to be wrong. Right? Therefore, in the wrongness, you immediately shatter the whole position that the ego has stood upon in defense of whatever you are defending. Only then can you see what you need to give up. Only then can you see what you have been defending, and therefore begin to work it out.

"There was another concept there: that when we see that ego come out and we are projecting it out onto another person, we start to look upon that projection as a valid justification to maintain a superiority as far as one Unariun to another Unariun is concerned. In other words, 'I have realizations that you do not have. I will keep them to myself but I am superior to you. Nevertheless, I won't tell you that I am superior to you, but my whole frequency toward you will convey my ideas to you.' Immediately we have set our own selves into that pecking order, ego-evaluating system that we had just evaluated ourselves out of. But by that very realization, in saying, 'I am realizing this other person's past but I am not going to tell him. I'll let him work it out himself because somebody told me to,' then we have this realization that we are superior and this other person is inferior and because, 'I have realizations that he doesn't have. Even though I was negative, I at least realized that I was negative. This person does not.'

"This is all very dangerous thought-work on our part. It is just like dancing and footwork: If you don't know how to dance, you get on the floor and before you know it your feet are all tangled up. If we don't know how to think, if we don't know how to reconstruct a positive evaluation from moment to moment and make that attunement to the Brothers on the Inner, it is very easy for us to become enmeshed in false justifications of why we exist. It is not going around pointing the finger at other fellow Unariun students. It is also not Unariun to go around and beat yourself to death, telling yourself that you are less than something that should exist.

"I believe all of us have gone through that. I have gone through that, and I'm sure each person has, too. If you are not doing something you feel is positive, you are fairly forced to do something that, in an illusionary way, seems positive, for the moment, just to justify your existence. Why must you do that? You must do that because the ego is so strong it cannot just have you exist as a positive, mental being. You have to do something that is ego-justified and ego-motivated, and among the peers of your group, which apparently, at this present time, are the Unariun students. To your associates, you have to appear to be doing something positive so they will respect you and evaluate your life existence, as your ego demands you to. And yet there are these constant incongruities and dichotomies in your own nature that are making it impossible for that ego to justify itself continually because the motivations are for self.

"And in Unarius those motivations and ego justifications are easily dismantled by the higher frequencies, and we are left bare and exposed more often than not. And in leaving ourselves open and exposed is like a gaping wound in our ego-consciousness. We have so meticulously built up these ego states of consciousness, our self-awarding systems and also our attempt for approval by our peer group, that we are mortified - mortally wounded, almost - if we are exposed by some frailty or some inconsistency in our personality make-up that other people can expose.

"The point of this discussion is that everyone is attempting this justification; therefore, he or she is seeing that there is not perfection there, and is attempting to wipe off the slag and just sort of implant it onto the first bystander. Pragmatically, or objectively speaking, that is condemning another person, or saying, 'I saw him in another lifetime when he was a 'rat fink',' or, 'He was a very negative person, and you know, he's really negative today,' and all these things.

"Now we are aware that when a person becomes positive, we are very happy for him, aren't we? Are we? How do you see that, Frank?"

Frank: "I would say yes, that we are happy."

Channel: "Okay, then we are very happy for him when he becomes positive, and yet there is a little bit of jealousy of him, too. Isn't there just a little bit? 'He's getting all the attention now and maybe I've been doing just as good as he has, or even better than he has. He has been really negative and he's getting all this attention now that he is doing better.' There's a little competition going on there. Well, a little competition doesn't hurt anyone if it is done in a healthy manner. So, what's a healthy manner? Do we know what a healthy manner is? No, not at the present time, we don't, because we haven't been learning at the present time, the objective side of analysis. We are attempting to do this in becoming a Unariun. We are attempting to justify our existence in the Infinite.

"The Infinite is secure. The Infinite knows that every wave form that you are, both physically and psychically, is part of it. It does not need proof. It just is. It's an impersonal energy force, interdimensionally expressing itself. And yet we are not secure with it because we are still learning this personalized form of the Infinite, and we constantly have to build up a brick wall against all outside intruders, to maintain some semblance of security that we don't really have.

"In other words, we are building a castle around a black hole. There is nothing there to build around except something that you want to discard. You really want to get rid of it but you are building these castle walls higher and higher and higher; actually you are not walling out the Infinite but you are walling yourself in from the Infinite. So your defenses have become your prison. Your defense mechanisms are binding you very closely to the concrete wall in balls and chains and keeping you from that freedom that you so earnestly desire. And freedom from what? Freedom from some other person's criticism? Freedom from financial want? Freedom from insecurity? What you want freedom from are the things that have pressured or forced you to do things that you might not want to do. Whether they are what we call good things or bad things, (positive or negative), we are forced to do things because there are two forces within our being - the superconscious and the subconscious; the subconscious, or our past, trying to gain influence over our lives. We have not really determined which force is going to take precedence within us as of yet. The fact that you constantly vascillate, proves this."

Michael: "Is it sometimes necessary to force yourself to come to the Center and do work?"

Channel: "Yes, and I was getting to that. Very good. We can say that this force is our guilt from our past, and yet we could not have incurred guilt unless we had contacted the positive side of life, in the face of some master; in the face of some deed that we might have done, and inadvertently felt good and satisfied that we did it. But all of those energies were going into us. In other words, we have built up this superconscious force that is now trying to exert itself over the negative force that we have been in, many more lifetimes than on the positive side. So, guilt is actually incurred by the superconscious, telling that person that he is doing wrong now; whereas, before when he was doing wrong, there was no other force to tell him that he was doing wrong. Now these accumulations of little positive deeds or little positive associations on the inner worlds and the outer worlds, etc., the very little minute things we might do such as: We might find a ten dollar bill and turn it in; well, that's a little positive deed. We might help a lady across the street or we might save a dog from getting hit by a car.

"Well, all those are nice and satisfying, 'I did a good thing! I saved a life, or I saved a hardship', etc. Well, that is accumulated life after life. Instead of killing the gladiator, you decided to let him live after you have maimed him. You cut off a couple of his arms, but you let him live. Well, before that, you would never have even thought of letting him live. You would have said, 'Death to him!' and 'Kill him!' But then you got this little voice saying, 'Don't kill him.' That's all you got. You didn't say, 'God or Jesus Christ is coming down to save you from the Four Horsemen of the Apocalypse.' All you heard was, 'Don't kill him.' so, 'Okay, I won't kill him; I will be nice today. I won't kill him today.'

"Well, that was way back in the Roman times, and way back in Alexander's time or whenever. That was a 'little' good deed, in comparison or relative to the things you did before. So all of those little accumulated acts built up this positive side of your nature. So now, this positive side of your nature is telling the negative side, 'Hey, you're doing wrong!'

And the negative side says, 'Hey, I have been doing this all my evolution. Don't tell me what to do!' Now you have this devil and angel type affair; one sitting on one shoulder, and one on the other shoulder. It's just an analogy, but is fairly close to what is going on within all of us, all the time. The average earth man, sitting out there in the world, goes through the same thing all the time but he doesn't recognize it. He calls it 'Jesus told me to do it.' 'The devil made me do it,' or whatever. But really it is pretty much the same thing. We call it the 'lower self'; we call it the 'higher self'; we call it the Brothers and we call it whatever; the positive forces exerting themselves against the negative forces or vice versa.

"So you are doing good things because the good side won out in that particular moment, but the negative always tries to equalize or balance that positive thing you did. So it is constantly tugging away at your positive deed to equalize itself. 'You had your turn; now it's my turn.' It's like, well, you went to the meeting tonight, so why don't you go to the show tomorrow? 'Why not? I went to the meeting, didn't I? I did a good thing there and now I can go to the show tonight and feel really free about it.' You know you should have done something else, and that's okay because that's a form of balancing the positive and the negative. It doesn't mean that the negative has to express itself, too. But there is this regularity, constantly keeping on the line of equilibrium. It's not that some days you don't feel as good as you do other days, it's a form of the positive and negative shifting biases in your psychic anatomy. What does count is that you maintain a regularity of consciousness along that undulating line. To be aware when you are under the negative side or under stress, and then when you are on the positive side, when you are 'on top of it', is the way to strive for because you have to restrain the positive side of your nature to not get hilariously funny or laughing. That only swings you back down in the cycle, to a very negative pit or depression. (We do not mean that happiness is undesirable.)

"So try to maintain the equilibrium of consciousness as it is very important to we Unariuns who are trying to recognize these many past, aberrated forms of consciousness caused from these negations."

Michael: "I think that I have seen everybody here except Uriel expressing both extremely positive and extremely negative. Now is what you are saying: If I feel really rotten, I would do something else than just sleep; if I feel really good and creative and want to paint a picture, and since I don't paint, I would not do that because I would be going off my line of equilibrium?"

Channel: "This is very true, with some modifications. This is very true in the sense that you are aware that you are negative that day when you get up. You are also aware that you feel very creative. So in that awareness, you are not necessarily stifling it, but you are directing it, with purpose, which keeps you in that forward motion of progress. You are being aware even on the negative side; so you have control over it. You have some say-so over it. You say, 'I'm really negative today. Well, why do I have to be negative?' At least you are making a choice there, whereas before, you were going around in a black pit with a hood around you and you were not aware that you were doing all those things. But when you are aware that you are doing all those things, you have some semblance of control over your life. You have chosen, 'I am not feeling on top today, so I am going to make a special effort to get up on the line.' Then you say, 'I feel like going out and building the Empire State Building,' or, 'I feel like going out and painting the side of the wall at the Center!'

"It is then that we should hold in our consciousness: to not go out and put a fire to the world because 'It's so diseased that we have to destroy it,' or the other way, 'We have to go out and hand out pamphlets. Here, take one. I don't know what this is, I can't even read. Take one anyway, you'll feel better just because you have it. Next week we are coming out with one that's all pictures!' You don't go to other extremes, and doing what I just described or something similar, is being really an extremist."

David Keymas: "I find that some days, when I am a little clearer in consciousness, I will study maybe three hours of real study and introspection and have realizations, and feel like I am understanding something. Then I will go to the Center and work hard, and I will feel like I am about to explode from the built-up energy. I feel like I have gotten so high that I have actually overstepped my boundaries. I become fearful that I have gotten so high, that I am going to take a fall. Is that what you mean?"

Channel: "No, that is not the case! What happens to you in this particular case (in case anybody else has that particular problem and I don't see why you should have that problem), it is a form of the lower self, trying to negate the progress you are making. It is your lower self, just like we said, 'I went to the meeting last night, so now I am going to the movies.' It's always these two counteracting forces. That lower self, that ego just doesn't want you to study that three hours to accomplish like that. I am not saying, (and let this not be misunderstood) and the Brothers are not bringing the concept that you have to watch out because you are progressing too much! That is not what we are saying at all. What we are saying is: Keep your consciousness continually regulated and being receptive to all positive inspiration; within your frame of consciousness, to continually be aware of where your thoughts are coming from. If you are studying for three hours and you are gaining by that three hours of study and truly being introspective and learning, there is no way that you could be going 'too high'. There is no way. It is that other counteractive force that is trying to tell you that you are doing too much. So if you do that every day - even against your own will - study that three hours and come down and do that whether you feel like it or not, you will begin to learn what we are speaking of. No matter what you are doing physically, it's where you are keeping your consciousness.

"Uriel has said this so many times but we haven't really learned it. It's where our consciousness is in whatever we are doing. It just so happens that on that particular day we were riding the positive side of that wave which we were talking about, but we were just letting it ride. We were not aware that we were on the positive side, 'even though I felt good today!' But that doesn't mean that you were aware of the energy principles of why you were feeling good today; nor are you aware of why you are feeling so rotten today. But the only time we really begin to analyze is when we do feel rotten? Right? Because we are fairly forced into doing it. This is why being forced into doing something is very necessary in the beginning stages of the cocoon metamorphizing into a butterfly. It is the pangs of birth that we are experiencing. These pangs are: being forced to do something we don't want to do. It's not somebody else forcing us into what to do; it is ourselves forcing us to do it. And you actually go around with a pained look on your face, like somebody is beating you across the back."

David Keymas: "That really helps a lot. I really appreciate that. One other problem I have, that you might be able to clear up - a misconcept: In that same day of feeling I am going too high, half that day, until about 4:00 p.m., I will be feeling really tremendous and good about myself and all those positive attributes, then right around 4:00 p.m., I feel a drop and I feel that I have to experience the other side of the wave. So when it comes, I just accept it saying, 'Well, you asked for it,' so I don't have to go down with it."

Channel: "No, not at all. And this is what Uriel means, 'Wake up!' Just don't ride that wave; be analytical, be objective. Why am I feeling good right now? Why do I feel a turn in my consciousness? Why? Why? Why? You are continually asking yourself, 'Why?'; taking the old thermometer and putting it in there. So, you are continually adjusting your consciousness to the Infinite flowing into your being. This is where we go wrong many times, where we are flat-out, hard rock to the flow of Infinite Intelligence to us. We are unyielding to the circumstances. Even within our own being we are unyielding; even if it's negative, we are going to it. What is negative? If we know it's negative to us, it is negative.

More often than not, we are unyielding."

Michael: "There is something bothering me right now. Yesterday I went to the dentist and he gave me a general anesthetic. I missed work this morning and am very incoherent in my consciousness right now."

Channel: "Just keep listening to the tapes."

Michael: "I'm lost, because in my own past I have been involved with drugs, and all those frequencies have come in. Is it possible to lose progress in a situation like this?"

Channel: "This might be a very good learning experience for you, but I wouldn't say that you lose progress by going through an experience. You are going through this experience, is all you should say, and you are going to learn from it. You are going to become stronger because of it. And I am going to be more of a staunch Light Bearer because of it. That's my purpose. Don't be concerned about this because many times the Brothers use this as a means to take the soul into the astral worlds and do much work on this person."

Michael: "I thought I was letting . . . "

Channel: "Well now, you are letting the old negative forces tell you that it is a negative thing. You are letting that negative self say, 'I want those negative drug frequencies there, yet I have such a guilt because I knew that it was wrongly done by me.' By beating yourself - and that is what you are doing right now - you are beating yourself and therefore drawing all those obsessions right to you! You are a good proof and example of what we just said. If you go around saying, 'Woe is me! Lost is me. No good is me, and I have no value, no worth. Why am I living? Why am I breathing and existing?' and all this and that, you are running around here like that, well, that is just as bad on the soul and it brings in just as much obsession, and realizing that I was a murderer - that's the worst thing we can think of, but there are many, many things that make us evaluate ourselves as negative people. And, 'I was the worst of the worst in the Unholy Six Wars! I have blown up planets and enslaved billions of people, and done all kinds of atrocious acts and deeds against them, and therefore, all of that is judged by me as unworthy of existence!'

"It is very hard to keep your perspective when you have all of that evaluation there; yet, if you aren't studying and trying to learn the principles all the time in conjunction with all of these horrendous experiences that you are becoming aware of (that the Brothers are helping us become aware of) we are going to become lost in this miasma of self-doubt, self-destruction and self-defacing. We're going to put ourselves in the proverbial gutter of life itself, which is totally ridiculous, and it is not the way of working out - not by me, not by you, and not by anybody who goes through this process. Everybody does it, but that is not how we are going to get out of it."

Jeff: "Isn't it some kind of an ego trip when you set some kind of artificial standards and then, to find out that you were all wrong, etc., the 'Oh my God,' and all?"

Channel: "Not 'my God', but, 'me God'. See, that's what I am saying in the sense that the ego is perfect. It is built and constructed on the basis that it is perfect, and there is no room for cracks or any kind of an incorrect situation within the ego. There is no room for any of that - 'No monkey business!' It is perfect, and squared off, and it is everything - so we think! So that means that we have got to be God! God is perfect and the ego is perfect, so therefore the ego has got to be God. Any time God is brought into play, you feel like, 'God and me are like buddies. He's a lot like me.' Well . . . . doesn't man give God all of His attributes?"

Jeff: "I have this picture of my ego going down in flames in identifying with it. That's why I don't get down when I find out I wasn't somebody I thought I was, because I am still identifying with that ego as it was dying, down in the pits."

Channel: " 'If you don't watch out!' ... You see, the problem is: If we don't watch out, we go down with our ego. This is the whole purpose of this discussion; we cannot go down with our ego as we are annihilating it, in the process of progression here. The ego has value only to a life that we are now trying to discard. If we are discarding this life without picking up a new type of evaluation, then we, too, will go down with this annihilation. In other words, we will be annihilated as beings if we don't place the Infinite within us in perspective; that there is value there and yet we have not touched that Superconscious Self in any way, as of this day, that can be considered in any way developed. Therefore, we just say, plain and simply, 'I'm in kindergarten!'

"Now you are in kindergarten, not because you don't belong there; you are in kindergarten to learn what kindergarten has to teach you. This is kindergarten; it isn't a home for the aged; it is a learning place. And most of us are learning that at one time, we thought we were pretty developed, but we have never been developed; never have we been called developed souls! This is an erroneous concept also, and this is why we got into so much trouble:

because we thought we were. You might say we were in first grade and thought we were in college. Now, can you imagine a first-grader walking around college and telling the professors, 'I'm in my fourth year here and I want you to teach me everything you know.' And the professor says, 'Yeah? You have third grade coloring books with you. How am I going to teach you anything with that? Go back to the first grade, and when you learn how to color red, you'll be back.' So this is what has happened to us; we have come a little way, and reverted; got sent back a couple of grades. But we have to keep our evolution in a perspective at all times - not our ego; because you might say, 'Aha! I've been down on myself too much, now I am going to build up my ego!' This is not the point of this discussion either."

Steve Sandlin: "The ego is down on itself?"

Channel: "Oh, no. It is not! You see there are many portions within your psychic anatomy that have self-judging evaluations within the total spectrum of your being. The ego is going down in flames, it has already been deflated. True, we must say that there is another form of the ego that takes over, but it is not the same type of ego that is being deflated. It's another defense mechanism. That is a portion of the ego - I'll say that."

Steve: "When you keep yourself down, you are keeping yourself in vogue with certain standards. It's like when you said we are in kindergarten. I immediately beat myself down to be in kindergarten, to keep in vogue with that standard you just set. Or for instance, I say that tomorrow I am going to work down at the Center for four hours and I only work for one hour; I spend the next two hours beating myself in vogue with being progressive."

Channel: "That's what I am saying you don't

have to do."

Steve: "What part of you then is beating yourself?"

Answer: "That is another portion of the ego."

Michael: "What you are saying is that no matter where we are in our own realization as far as spiritual progression goes, we are in kindergarten, and no matter what we think about ourselves as souls, we are still learning those kindergarten experiences?"

Channel: "Very essential, though what we are saying is: that no matter where you think you are, you have to find out where you really are to become healthy. We all want to become healthy, right? I tell you the Seven Centers of Unarius are heaven to us, but where we will still be in kindergarten because there will be a heaven above that, and a heaven above that, etc., etc. You will always feel that you are in kindergarten if you are healthy. Because if you are healthy, you know that the Infinite is infinite, and therefore no matter where you are, you are in kindergarten. But that comes only with the maturity and advancement of the soul - to find out his state of being at any one particular moment in the Infinite Consciousness; in the mainstream of consciousness.

"If he is aware that the Infinite is infinite, and that he, himself, is an infinite being no matter where he is, no matter what level he is on, he is not judging the ones above him and below him. He is what he is, in the Infinite Intelligence; he is a healthy, progressive being at that moment. What we are trying to learn to be, is a balanced, healthy being. Well we have to realize that we have been negative; we have to realize that we have been a little positive - and that is what we are striving to be - positive! We are neither black nor white. I was talking to Stephen today about this very same concept; that we are in the process of sifting apart the greys of our consciousness into black and white: What is of Spirit, what is progressive to us and what is negative and regressive or the chaff of the wheat? The wheat is usable; the chaff will only gag in our mouths, so we discard it. It doesn't mean the chaff is some evil, gory, horrible monster; it is just unusable now, so it has been separated from the usable.

"Now, that's what we are trying to do: We are trying to separate the things that are no longer viable to us, from the things that are usable to us in some future world where we want to go; a world we know exists and where we have put in our reservations. We know we want to go there; we know that it is our eventual destination if we put in the proper prerequisites which are overcoming the self; sifting all of those past evaluations - the usable from the unusable, then discarding the unusable and taking the usable as food and sustenance to the world that will sustain us. That food is from this world but it has to be sifted out from the slag - the unusable."

Patricia: "Sometimes we see ourselves as having been a horrible monster, and some part of us that behaves like a monster. It is necessary that we see them as a monster before we become objective with them and say, 'That was a part of the Infinite expressing itself?' Must we see them as a horrible monster first?"

Answer: "Yes, because that is how you generated it: as a horrible monster. So therefore, you have to reincur the experience of how you generated it. So this is a very good question and the answer should apply to all of us in the sense that: We can very easily say that horrible monster is still in force, even after we recognize it and we must therefore use it to beat us, and that is what we are doing. We are using that monster to beat ourselves down, into submission. Remember how the monks used to beat themselves and flail themselves, and used to go around hitting boards in their faces? It is very true; I read into that after I heard it from one of the students. They used to flail themselves silly. It's because they are beating themselves out of what they feel is a monster they have uncovered in their own nature.

"Other people are sitting on beds of nails and all those things, putting skewers through their bodies; then the beating of the self is actually a way to justify that monster. That monster is there because, 'Now I have built it up and I have to go to see the other side of the wave, so I will let it beat me down into a pulp.' My evaluation of myself will slowly sink down into the horizon, and there is nothing to replace it. You people haven't really replaced it, you are still a bunch of gossips. You are still a bunch of people, telling all the other students how much you don't like this person; how much that person is such a nasty person, 'I don't like to be around that person and I don't like to be around this person.' You are just a bunch of teeter-totters. That's the truth.

"You think you are being objective but you are not. You are only transferring all of your insecurities onto another person, whoever he is or whatever he is. If you are not telling him, you are telling other people, 'Well, look at that nasty person! Do you know what he's reliving?' etc., etc. There is this little gossip. And it is true. I hear it all the time in corners until I come and they stop talking. Then I pretend to go out but am just around the corner and everybody goes back to talking again about that very same thing. I am amazed! Here we are, objective people, acting like little packrats in little gossip holes. That's not the way to be a Unariun - to beat the other guy down! Just because you aren't telling that guy to his face, do you think he is not getting those energies? That's not being a help to him by saying, 'Oh, cootie-cootie! Look what he is!'

"Now, I'm sure that in everybody's true self, in your true nature, you would not have it that way; that each and every one of you, in your soulic self, really and truly love each other and would do everything in your power to help one another. I have seen that too but what we are saying is: There are two sides to each of us. We are not all black. It reminds me of the tar pits in Los Angeles. Did you ever go to the tar pits in Los Angeles? Well, you could look in them and it was all gooky and icky. Some of us get to the point where we think we're all black and no good, and then others of us take on this, 'Oh, I'm so positive, I don't know what to do with myself. I just want to teach everybody. I just want everybody to know just how positively positive I can be and I want to express it to everybody.' And yet in the dark depths of this soul there are these ugly little thoughts; ugly little monsters, clipping at this person's heels.

"What we are trying to say is that both sides of your nature are trying to grasp and take hold of you during this particular point in your evolution. Please never lose sight of the fact that there are these two sides to you: There is the more positive nature, attempting to lift you, to transcend you, to radiate you into a world that will sustain your life into an eternal and loving nature; then there's this other side that is continually trying to propagate what you have been - the ego structure that refuses to submit to any other opinion, and always has to be superior or even has to be inferior. Constantly, it must justify its existence through any means it can because its existence was doomed from the very start. From the very day it was built, it was doomed one way or another; either to be replaced by a spiritual butterfly from the cocoon or by its own tearing and rending away of itself to eventual annihilation and reabsorption. So you have both these selves vying for power within your being.

"Now we have this continual reliving going on, vying for power. One day somebody's top dog, one day he is low dog; or 'I am third dog or fourth dog. I have a list and I check off which dog I am today.' Isn't this true with yourselves? Where am I today? 'Not in here, (speaking of 'within the self') but out there.' And you don't really stop to think, 'Where am I, inside myself? Who do I really have to impress? Well, I know I am doing a lot of wrong things so I have to impress others to keep them from seeing that I am doing these wrong things, or from even suspecting that I do anything wrong.' I don't know what you are doing wrong; you may be going out and painting all the fire hydrants green at night, but you know what you are doing wrong; you know what you have evaluated as wrong. It is not what I have, or what George or Jane has, but it is what you have, yourself.

"In your own position, you have a set of values and those values are kept by you or reneged upon by you and by nobody else. As Unariun students, we are moving along with certain group standards because we all have complied within our own natures that the 'Unarius

Principles are what I understand to be what is necessary, for me to learn to become a healthy soul; therefore I am a part of this group or I am a Unariun.' Now, all of the personal little prejudices and interactions we might have amongst ourselves are a part of that personal evaluation that we have developed throughout our past, interacting with each other, so all of that is part of the group consciousness: 'We have been negative in the past, and we are attempting to rectify it at the present time, into a positive status toward the Infinite, individually, and of course, collectively because we have all sorts of intertwining with each other in this so-called power struggle.'

"Now the power struggle is only within the self because a master doesn't care if his job is sweeping up at night. In fact, he would prefer that. But I will tell you, among the average earth men, there is a lot of grumbling going on because he has placed himself as a lesser person than the guy who turns the key in the morning. So he has placed himself in that pecking order instead of evaluating his own soulic evolution from internal sources because he doesn't have that true evaluation as yet. He is still hiding much of that anger, resentment and insecurity within himself and he doesn't have the proper balance that has that thermometer out all the time, within himself. He is continually evaluating his worth by how others see him. Of course, that is in psychological books and what we are saying tonight is very apparent.

"Yet, we extend these psychological concepts into the realm of the psychic anatomy, into the realm of how energy operates in its inviolate principles. In doing this, we can gain some foothold into the awareness of ourselves so that we no longer need to go around with these gossipy types of frequencies and consciousness, and reading for other people, beating ourselves, or elevating ourselves into some evangelistic type of emancipator that some of us might feel that we have become if we have received some compliment that is either deserved or isn't deserved by us. You see, we are always going to extremes; either we are no good, we are black; or we are perfect, we are white! This is because of the Battle of Armageddon. We have ceased to be grey in our consciousness. We have ceased to be just ordinary folks out there - neither good nor bad; neither positive nor negative, neither hero or coward. Most of those people are that way. Now, they will all change and they are changing. We have all changed and we are changing.

"We are speaking to this little group at the moment, yet these are Universal Principles that anybody can prove to himself. You don't need anybody to get up here to say, 'I'm going to give you a new set of laws.' Well, there isn't a new set of laws. The Infinite has been infinite long before anybody called himself a name. So we have started to sift the chaff from the wheat and now we have gotten into the grey matter and we are separating all those things that are negative to us, personally, and positive to us, personally. In that separation, there is much more pressure because we are seeing the good and evil within ourselves - or the positive and negative - so much clearer than we ever have, that it is becoming very clear to us that we have been very powerfully negative in the past because there is not very much wheat and there is a lot of chaff. There is a big pile of chaff and only a few little grains of wheat. So, what happened to all the good things? While it was in the grey area, we just saw a big pile of experiences that neither had any meaning or shape or form. It was just there and we were living, breathing and eating and doing all of the good things that this good life gives you. And then we started sifting them apart, and we have just a few golden grains or pure seeds of wheat there but we do have a lot of husks.

"Now our problem is: 'How do we balance that all out. How do we?' We must realize that all that we are discarding was at one time very necessary to us to grow those little seeds, or to ripen them, or to bring them to the point where they could be consumed in the world where they originated. But they have been all buried under. We needed all this fertilizer here to gestate them. We needed all these negative experiences to create a sod for them. They were very necessary. Now they are not necessary, so there is no need to beat, but there is need to recognize; there is a big difference. It is a real monster and it actually is what you are now. You have to see the monster as it is. When you see it, you are that monster, momentarily. You _are_ that monster; you really are, and I know! But with recognition, the monster evaporates. A form of it can still come back really easily if we cry, 'Oh, the monster! I have really been horrible', and you go around with that state of consciousness continually, day after day. That monster will reemerge in a different form to devour your progressive chances. And I know this for a fact.

"Any questions?"

Steve: "After doing this for the thousandth time and having generated this opposite monster, how do we start to tear that one down?"

Answer: "It's like two dinosaurs in battle. Did you ever see the movie "One Million B.C." where this man is standing in the cave and these two monsters are flipping all over the place and their tails are beating down right next to him; the ground is shaking and it's 'earthquake' all over. Well, you can use that analogy of these two monsters fighting with each other, and you are that little guy, waiting for the outcome. But this is how we see ourselves - as very little importance compared to the two monsters that we built; it's the old Frankenstein thing: 'I can't get control over this. They are going to fight to the death and then the one that wins is going to eat me. I'm just a little morsel to be eaten.' But it's that one monster, fighting with the other monster. It's those two monsters fighting, and they are kept so busy that you are shivering over the outcome. You are the battleground for these many forces that have been built up by you. We have built up this set of monsters out of thinking, 'That's the way to survive!' But it's almost like we outlived the need to survive, in their eyes! We've sustained them for a long, long time and now they have outgrown us - let's put it that way. They have actually become so strong that we have no control over them."

Brian: "What you are saying then, to me means a lot. I have given up the idea for what I was here, to begin with, when I say, 'The monsters have got me; they've got me. Oh, that is bad, negative!' What I have forgotten is that it is I! So 'they' have not got me, because I am the only one who has the ability to make the decision . . . "

Answer: "Let me stop you there when you say 'They have got me,' or 'They haven't got me.' Let us put it this way: As long as they exist as a force, they are you. No, they don't have you, for they are a very part of you! You can't get rid of them until you, yourself, change them. You, yourself, have to pull the fangs; you have to get rid of the fingernails and claws. You have to tame the shrews within you, and it isn't a matter of being outside of you! I know there are a lot of misconceptions there also, about that. 'These obsessions have really got me today!' and all this. 'Oooh, ick, ick! Obsessions, obsessions!' as something outside of yourself. Those obsessions are you and nobody else! The only reason all these other hangers on, as we call outside obsessive entities or obsessions can even hook themselves to you, is through frequency relationship; that you, yourself requested and drew them to you, only as a result of what's inside you already and that <u>you</u> have created."

<u>Brian</u>: "It was just the idea that I was thinking of it differently. Maybe my higher self is only so big, but I'm doing it for myself. I forget that I have a good side to my nature. I don't want to be the negative side but the negative side is much more powerful."

<u>Channel</u>: "You have to be very careful, and that is what we are saying tonight. In balancing the scales over to that negative side, we don't have the directive balance or the integration yet to not let it get to us to such a point that we use it to sustain the very same thing that we are trying to get rid of. It's another form of hanging onto the past. It's another form of it."

<u>Dr. Gary</u>: "First of all, I have nothing prepared. I am not a member and I am very glad you have allowed my presence. I have a few questions. It appears that I have been wrongly educated, so I would like some things reconciled. I heard it said that the ego will die immediately."

<u>Answer</u>: "Not the Ego, just a portion that is exerting its influence at that particular

moment. It's such an enormous thing."

Dr. Gary: "I have been thinking all the time that the ego never really perishes or dies but it is the measure of withdrawal of the self called the ego that puts in this condition, this adverse condition, when we withdraw from the ego rather than the ego perishing at all. Now for my own education, I would like you to clear this up because I think it is very important whether the ego is at all vital or whether it could be reduced, or whether there is a god part of that self. Can that be reduced or can that always be the part that puts in that condition?"

Answer: "Like we have said before and maybe we should put it in different terms: The ego has expressed from the moment of our soul's conception and has been very, very necessary. Everything that we have built up, as we have referred it to the ego or the survival self - and that's really what the ego is, a survival self (and that's what we are talking about), that perfect-seeking expression, the ego self. It is very necessary to survive in the physical dimension with an ego, because it gives us that impetus or that drive. Now we can even say that the ego is part of the Infinite because there is nothing that isn't part of the Infinite. That's a part of god, a part of the Infinite.

"Now that self, after arriving at a certain point in its evolution, that ego ceases to be positive and progressive in its motivations and reverses to become what we refer to as the id, which is now beginning to be destructive to the person's progression into the Spiritual Worlds where the ego, as a survival self, exists. It does not exist only in the material world as a survival self and is a very important part of the person's evolution. It's important."

Dr. Gary: "I have another question. I think I was here when someone said, 'My God!' and you said, 'Me god'. That to me implies that we are God. Then I would like to ask for some sort of explanation there."

Answer: "I know what you mean. That is what we just said, that: There is nothing that is not a part of god. There is nothing; every atom, every universe, every little single insect and every being is the Infinite or God."

Dr. Gary:"Therefore, God manifests as plants and animals. I have not been accepting that within my being there are any parts of monsters. Maybe I'm wrong."

Channel: "Not parts, but whole monsters. I think that what we are trying to do is to dismantle them. I don't mean to be funny with it because it is very serious that a person becomes aware that he has done and expressed from his past lives, things that he thought were positive then and are, at this present time, now regressive to his spiritual evolution. That is why we say that all experiences a person goes through are not necessarily evil while he is doing them. If he is committing a murder, or defending his country, or whatever he's doing, if he feels it is right, in that moment of his evolution he is not, in his mind, creating an evil thing. So, later on, that very same experience, many lifetimes after, he has learned that killing, in any sense of the word, is not the progressive way, then that entity in his psychic body becomes evil to him and he must neutralize this energy. We can refer to it as a monster within him that needs neutralizing. That's why we are talking about monsters."

Dr. Gary: "My impression is . . . well, I haven't met anyone who has come back from another life and that past, having done something positive in the past, but through all my life, I have heard this talk of being monstrous. I have been taught that positively, all around me is Light which is positive and there are no negatives at all. If we are part of God, it is then, that what we think, prevents us from communication with people and God?"

<u>Answer:</u> "That's very true."

<u>Dr. Gary</u>: "Therefore, it is very necessary for us to constantly think of the positive side and to enhance this by doing things right, with charity, and by our own actions, good or bad, we mold the kind of person we are. Now is it right for us to think that we had monsters within us? As a matter of education now?"

<u>Answer</u>: "Well, it's not a matter of whether you think you have monsters within you, it's a matter of fact. If you do have various emotions that you no longer particularly care for, there might be drives that you don't particularly care for or there might be things that you have done even in this life that you are not particularly proud of - these things serve to add to the substance that there is an unhealthy malignancy in your being. Now let us go another step further. A person learns to be positive; this is how energy operates. It is a learned thing, it is not given to you. You say, 'I see Light all around me.' Where did you learn that, except through experiencing non-Light. You must, sometime in your evolution, have been in total darkness to recognize total Light. So to be in a total darkness area and to have that recognition, you must have already experienced all of the opposite to that Light - to that positivism; which is the negative side.

"So therefore, if energy operates inviolately

- which it does - then this negative energy must somehow be attempting to express itself as a polarity reversal within your own being and it must be that it is endeavoring to express itself outwardly into your life. Now, to compensate for your guilts, through your attempts to placate that negative side of you which you no longer want to look at because you want to be positive, you want to be a good person, you must strive as exceedingly well as you can to do everything as good as you can to negate that which you do not like, which is the non-Light within your own being. There are opposing forces within the Infinite, on extreme sides. The balancing factors of nature apply not only within the galaxies or the solar systems to keep planets revolving around their suns, or atoms in their proper orbits in a molecular combination, but also the forces that keep you as a being, sustained in your daily acts of life, must also have a positive and negative phase reversal polarity pattern so that you can live, you can breathe, and you can exist. It is the relativity of these negative and positive balancing forces in any particular point in your evolution that determine the forces of the absence of Light or the forces of Light that are influencing your life.

"Now you learned in your life from some other source - and it sounded good to you because you could escape your negative thoughts - that there is no such thing as negativity in you. Therefore, you could just forget about it; that sounded good to you, and it sounds good to a lot of people, but unfortunately, they are waylaid by this false conception because they are avoiding and escaping that which they must face one day; that is: that they have lived negative lifetimes. They have lived negative experiences, and as a matter of fact, they are living negative experiences today that in some future lifetime they will deem as negative, even though it might be positive to them today.

"So it is a factor of awareness, of realizing how energy operates within all beings, within all forms and structures within the Infinite Cosmos, interdimensionally speaking. Energy does have a positive and negative phase reversal for it to oscillate and for it to express. If you only had one side of your nature, you would be denying half of the Infinite."

Dr. Gary: "I was under the impression that we were born clean."

Answer:(A facetious reply from the Channel was given.) "Well, that depends on what hospital you are from!"

Dr. Gary: "Children, when young, seem uninhibited."

Answer: "Mentally, they are not inhibited or supressed by inhibitions and/or insecurities, and all, in their subconscious."

Dr. Gary: "When children are playing with other children, they have something which they lose when they grow up. We bear hostilities. Now if two people were born at the same time, and they had the same parents and they grow up together, and one is a good person and one is a bad person, this must be attributed to the fact that one has assimilated the negative conditions, while the other has assimilated the positive conditions. It is difficult to put the differentiations. They were born of the same parents and at the same time, but are totally different. How would you explain that?"

Answer: "That is a very good question and it is a question the answer to which would put man a thousand years in advance to where he is now, if he would conceive it because the answer that man seeks is: 'Where do I come from and where am I going?' And that was basically your question, 'Why does a person do what he does under any and all circumstances?' It is because a person is an evolutionary being, first, last and always. He is a spiritual being; he is not only a physical body. He takes into himself and he imbues and impinges all of the experiences within himself; all of the experiences from not only one lifetime but from many, many hundreds and thousands of lifetimes, and he carries these experiences over from one lifetime to the next.

"However, he does this within the constituents, with the demodulating nature of energy, to build himself a brand-new, sparkling subconscious mind that enables him to live each life in the idiom of that time. That subconscious is constructed partially out of the past experiences of this person's past lifetimes and his environment. These are the catalytic agents that mold and shape a person's outlook, personality, characteristic behaviorisms, etc. All of these things are outgrowths and outpicturings of this person's internal association with past experiences and the loved ones around him. They don't necessarily have to be his siblings or his parents. They could be friends, they could be influential relatives, uncles, aunts and persons of these relationships.

"Well, he carries these daily acts in what we call the 'subconscious', and yet the subconscious is continually associating this person with all of his past-life experiences. In this combining and recombining, it reshapes and changes this person's outlook on life in his daily acts, deeds and associations, and yet they remain essentially the same, lifetime after lifetime, until the person, himself, after he dies, seeks a higher intelligence than his own to find out the reasons for his being in existence. Therefore, there then come to him, many of these angelic beings; these Higher Intellects who begin to teach him slowly and arduously how he can live in a higher world; why he exists; where he has come from and where he is going - all of the answers to life, itself. Yet this person cannot learn overnight so he lives many, many other lifetimes, experiencing all of the happiness, the joys, the heartbreaks, the pain, grief and the sorrows and the sufferings of the physical life.

"In each lifetime, he seeks a little more wisdom, a little more knowledge in these higher planes of development, and later they become more and more real to him; so much so, that when he reincarnates back into the physical world, he seeks to learn or relearn more of these higher worlds while in the physical world. They are so real to him now, that he seeks to bring these higher worlds and their reality into his consciousness in the physical world. Therefore, he is looked upon as a wise person or a good person; he is looked upon as a person who wishes to build a beautiful society, or he wishes to paint great art works, or be a great scientist and discover the laws of nature. All of these are reasons for him being the way he is because he saw the reality of a higher, more perfectly-constructed dimension or spiritual world.

"He is getting his inspiration, his bountiful knowledge and his intuition from what he has learned on the higher, inner Worlds of Light. Now, we can say that another person who is also incarnating on the physical worlds, has not arrived at that point yet, where he is coming and going in his sleep state, his transcended state, and his life in between lives, in these higher worlds; who is yet suffering in a subastral world when he dies because he does not have the knowledge nor does he wish to have this knowledge impinged into him because of his long-suffering ego. His long-suffering state of mind wishes to relive and reinstate the subconscious malignancies that he has incurred in various destructive or negative past lifetime cycles. Therefore, when he is incarnated, he attaches himself to the same parents because these parents have had associations at different times, in different cycles, with the two different people in varying relationships. Whereas the so-called 'good person' might have been the father to one of these parents and was 'good' to them, it could also have been that the negative person or the not-so-positively developed person might have been an evil overlord to the parents. Through frequency relationship, he is attracted to the parents. So, both of them are attracted but for different reasons. By frequency relationship and energy patterns, we can see this in every family. All of the negative and positive associations have been meticulously knitted and built by these past circumstances - these past life experiences and lifetime after lifetime these groups of circumstances under different experiences have recurred so they can learn of life in this type of association.

"But it still boils down to: 'Does this person wish to be a progressive person? Does he wish to know and have the awareness that he is a being in the Infinite Intelligence?' The person who is coming and going in the spiritual worlds says, 'Yes, I am a spiritual person and I will seek to become a more infinite, spiritual person.' Then he is on the road to emancipation out of the physical world. It might take him a million years, but eventually, giving him the benefit of the doubt, he can progress into this higher state of awareness, whereas this person who has not developed is still languishing in the negative aspects of his own nature and it is projecting him outwardly to say that, 'The world is evil!' or 'The world owes me a living. My mother and father are no good and they don't treat me right.' All of these experiences are a part of his own inner growth or non-growth, whichever the case may be.

"No two people are on the same evolutionary level; therefore, you cannot expect two people to act or express emotionally the same or have the same personal characteristics or thought trains the other person does. We could go on and on all night with these concepts."

Dr. Gary: "I have been here only a few times; most others have been here more. In relation to the thought of accepting that you have monsters . . ."

Answer: "Everybody who lives a physical life has these negative experiences and we might have named them monsters. But scientifically speaking, they are past experiences of a more negative nature, to the positive quotients of energy that we wish now to express. People who experience these expressions from their past will call them monsters (little green men). I have seen snakes come at me because of my past forming of these negative deeds. There is no one, as I said, who becomes a Spiritual Being who does not go through the experience of not being a Spiritual Being. So, if you want to call them negative experiences, or learning experiences, or you wish to call them monsters or astral influences, astral forces, whatever you wish to call them, they do exist within every evolutionary being who has lived and who is living on a physical world. It is the way and means to experience the Infinite. First, we must experience the negative before we can experience the positive side of the Infinite or the finite expression. That doesn't mean that everything is a monster, but if we allow it to control us, then it does truly become a Frankenstein monster. And those are very strong words and sometimes we need strong words; better yet, sometimes we need even stronger proofs within ourselves.

"I think we will discontinue for now."

* * *

ILLUMINATION

Out of the muddy swamp I was in, I felt around my grave for some light.

So here I was, peering from out the darkened gloom and into what seemed to me to be a light.

"What is this Light?" I asked. "From whence did it come?"

And this question persisted and nothing else my mind did hold.

As I gazed into the Light, the closer it became until I and my surroundings were illumined.

I had not known where I had been and as I looked, my eyes beheld in shock, a cabin pitiful in comforts.

But why have I been here? I asked this Light and it spoke thusly:

"Thou hast been a soul, who in thy evolution has fallen. Ye took the pathway that hast led to this - where ye now standest.

Thou hast rebelled against the Light of Truth. Ye must come now to the parting of these ways and now take the upward pathway.
And dost not thou feel this to be true?"

As I looked upon this Light that was all but blinding, I felt Love, Compassion and warmth, and from my eyes the tears did fall.
For the Love was just too great!

"Thou hast, in thy judgment, placed thy self here. Now is the time of forgiving thy self for deeds done so long ago, that any soul would not be proud. Thou must come now."

The Light beckoned to me and never did I feel so much gratitude for being rescued from such concealment. Hiding away in pity and shame I was, that never seemed to go away.

I followed this Angel and was told many things of the past, present and of the future - how life would be.

The plan I learned only by degrees for so vast it is to conceive. My place, too, was explained to me
What change I was to undergo so that my heritage would be as She - a Sister of Light.

I had begged much for this opportunity so it has been granted to me.

I found my place among earthlings, in a world just like a world I had come from - unillumined and dark that needed too to be rescued from its reckless plight, unto the Light of Spirit and Truth.

So it was, I was to come and be a part of a world-changing Mission
The last chance given to the dark ones - the rascals of evolution
To make this their last stay amongst a

rhetorical, material vibration.

And my heart rejoiced, for in the future I saw the plan successful

And as I rejoiced, countless others joined in too, until the whole world rejoiced in its new-found Light

No longer a bad apple amongst others more progressed

No longer darkened countenances caught up day-to-day, being frantic materialists

But peacefully knowing, too, that there is a Universal Mind of Oneness and Intelligence

And that in the asking, would be the receiving.

So, rejoice as we rejoice in the successful efforts of an Archangel

Knowing that our self-punishment and persecution by ourselves, is over.

To know TRUTH, though it be the bitter dregs of our past is far better than the darkened silence of self-defeating ignorance.

For is it not ignorance that eventually leads to Death?

- Margie Vasquez

* * *

Letters from Students Regarding Their Past Life Flashbacks and Overcomings!

May, 1979

Anne: As I have not given any testimonials yet about the realizations and workouts I have had with this tremendous Unholy Six Cycle, I would like to tell you about them here, to polarize them with yourself and the Brothers.

I had some realizations Wednesday, May 30, listening to the testimonials of the other students, and I was also greatly helped by being a part of the Severinus contact for the video. As always, the personality that I had at that time came through very strongly. In the contact, I reenacted the part of being an assistant to a great scientist and I was a very hostile, superior person. I was just a 'yes' man to this scientist, which is something I have always reacted to in this lifetime. Also for another part of the video, we reenacted a murder and I was the one who was hiding in the bushes waiting for this person to walk by. I know that I have had assassin lifetimes, but mostly it was the energy of being a back-stabber or hiding my true intent, which struck me. I found it very easy to play this part, to be a very evil, destructive person and I was very glad of the opportunity to get this out and look at it and turn it around by being a part of the contact.

The student who was this scientist for Severinus (Hugh), gave a testimonial about plastic surgery and I really had that awful feeling as he was talking, so I know that is what I was doing in this particular lifetime during the Unholy Six. I have been very involved in the makeup for these video productions. During this cycle I have had to force myself to do the makeup and I have not wanted to study it. Many times

I become nauseous when I am doing makeup on a student lately. This shows me the horrible past I have had with plastic surgery and brain operations. I had the realization that I was one of those who did plastic surgery on those who volunteered to be duplicates of Dalos. I also reacted to Jeff's testimonial about the Pleasure Planet. I know that I did operations on people there to enhance their sexual drives, to just turn them into sex machines, and I believe that in another lifetime I was one of those who had those operations out of guilt.

When the Dalos cycle came in, I was feeling very blocked off. I let myself get negative and didn't study, and so cheated myself out of a lot of help and realizations that I could have had. But I do know I was in the procession to welcome Dalos and that I saw him while he was in the force field. I may have been somehow involved with the creation or operation of this force field, as I relate that to the silver mesh that I was going to wrap around the flower hoops. I had an image in my mind of a double helix of silver going around those hoops, surrounding them. Also while in the warehouse working on a light box for a console for the Basis contact, I was greatly bothered by loud noise and strong smoke from another student sawing some wood, which I relate to the torturing of Dalos. The idea that I reacted to most about Dalos' imprisonment was when I heard that he was kept in his own excrement for years. So I may have actually been one of the ones responsible for that dastardly negation. I know I was a scientist and a surgeon, so I could have been one of those at the controls, or in other ways to blame.

A really big realization for me was when I admitted to myself that I had some position of authority in this lifetime in the Unholy Six when we captured Dalos. For the reenactment of Dalos' reception on Orion, I had been invited to be in the procession, but I could not because I had to work. I later used this as an excuse to believe that I had not really been a negative leader then. But now we are to have another procession for the presentation of the Cosmic Flame Generator, and I was again invited to play a part. I could not hide anymore! So I had to face that past. I also faced myself; that I did not want to be in the processions; that I was trying to avoid a workout. When I realized that, then I stopped trying to fight it.

Shortly after the Dalos cycle was brought in, I found myself in a different lifetime, which I believe was a regeneration of that lifetime in the Unholy Six. In that lifetime, I think it was in the Middle Ages here on Earth. I had just started a new job but I was having to force myself to go to work. One day when it was time to get ready to go to work, I stood in front of my closet and looked at my uniform and cried and cried. I was terrified to go to work. I had the feeling, very strong, that some people would soon be coming to the door and would drag me off to this horrible prison, which was the restaurant where I worked. I had the feeling that I would go there and never come out again. Many things tuned me in to the restaurant as being a prison. First of all, my boss had me at his beck and call; he would call me in to work whenever he needed me. The air conditioner never works and it gets so hot and smoky, sometimes I feel like I can't breathe. I know this was a terrible prison and my boss was the head jailer. I believe I was a soldier because of the uniform, and that my function was to go to people's houses and drag them off to this prison, and they would never come out. It was a prison for the political dissidents whom the government wanted to get information from. It was a big stone place, very filthy inside (which is a feeling I always have about the restaurant, even though I know it's not), with a basement at the bottom which was where people were tortured. When I had this big workout, I had the realization that if I could be in on torturing people, I could easily torture Dalos, the Higher Being. I brought others in to be tortured, so I must have had a part with Dalos. That's why the reaction to dirtiness; to the smokiness, the noise.

I am sure that I will have much more later. Thank you for your help and your love. I am proud to be your student.

Christine: While sitting in quietude at home after the meeting Sunday night, I saw the Great Being that you are, Uriel, come into my consciousness. You were swirling and twirling and emanating great spiralled arms of force out from you and back into you, yet somewhat taking on a physical form of a beautiful Higher Being, like your Archangel picture that is here in the Center to tune in to you. I also saw my lower self exposed and how I have put great effort forth to destroy the many wondrous and incredible works that you have brought throughout the eons. Many thousands of years I have attempted to drag down your image of beauty and the Infinite which you are, and have obliterated my life in so doing.

The other night you spoke to me and said words such as, "Don't you like to ham it up?" meaning the acting we are doing for the 32 planets. I said, "Yes, it's fun." I didn't take into myself the true meaning or at least what I had to learn from the inner meaning which you are always speaking of. I saw how, over the thousands of years of the Unholy Six, Lemuria, Atlantis, etc., and to the present, I have tried to portray myself as knowing something of Infinite Consciousness, but it has been an act-up job; a ham-up job, just pretending to everybody a facade of lies to my fellowman and to myself. Such a sick one I am! But for this knowledge I am very grateful; grateful that you have revealed to me once again my lower self and in such a beautiful way, as you always do. I am awed at your great intelligence and see the very opposite that I am. But I am trying to change the old ways.

I am aware that the five years I have been privileged to be in your Presence is the greatest thing that could ever happen to me - what you are doing for us. *You are saving us all!* It is beyond me that it is really happening; yet it is happening! It is the wonder of wonders, and I am beginning to see life as it is truly meant to be seen, a place of love. This love exists within me now, an emanation of your Essence, the Infinite, which encompasses all and is within the hearts of all.

Nellie Gonos: Dearest Uriel, I must let you know that as soon as I read the note and before I read the dissertation, I knew it had to be unpleasant and my thoughts were, "What kind of drastic acts had I committed now?" and then I cried. You really opened up my Pandora's box for me when you told me the letter opener I sent you represented a very sharp instrument used in the Orion negative cycle. Reading the dissertation, then when I got to the paragraph of the computer world and people being grown and used for fossil fuel and foodstuffs, I reacted and broke down and cried. That was it.

Then the 7th day of reading, Sunday, while lying in bed scratching my discolored scaly skin, this flashed through my mind, "Dalos' skin was malformed. I did it with those electronic wave forms." I cried but I couldn't accept it then and there because I could not see how I was a political administrator. Then the 10th day of reading, I read slowly to absorb especially, the paragraph on administrators being political scientists (in our day they are politicians). Then it hit me! I never liked politicians . . . government. I felt they were dishonest, etc. In fact, I never registered to vote for them. Then I realized why I had this awful feeling towards them, because I saw myself as them and being one working for the government, etc. Now I am able to accept being a political administrator on the Orion world and could now accept being the one who used electronics on the Ambassador, Dalos. This was powerful to me . . . more tears. Shortly after, a very warm feeling came over me on the 14th day of reading. I must mention this and you will find it out in reading.

I listen to the tape *Psychodynamics of Self*, morning and night and most of the times I was taken out. This night I was taken out just before Lao-tse spoke and came back just before the end of the tape. Shortly after, I fell asleep until 3:30 a.m. Couldn't sleep for my arms and legs ached. Took several pills but could not get to sleep. Thoughts were racing through my mind, "No wonder you are suffering; you brought this all on yourself when you were cutting up all those bodies, especially at the joints, and taking great pride. It was as though you were carving a statue." Gosh, this is unbelievable! Then I heard mentally our dear Brother Lao-tse's words: "Tis so." With this, I really shed tears. A short while later I had a warm feeling come over me.

Before reading the dissertation on the 15th day, I've been pondering over why I had these blocks, mentally and physically. Soon found out after reading. In this lifetime, during my high school years, I hated science, physics, algebra, biology, astronomy and never knew why because when on Orion, I never took interest in the galactic structures, in structures concerning my own life source. I was concerned only with the needs of life, material-wise and could manipulate very well as I do now with finances, taking from one account to put into another, etc. Knowing and feeling this way did make me feel like a god. What else! Also when my husband and I quarreled, I always said to him, "Who do you think you are . . . a god, that I have to run to your beck and call?" This is really a reliving of the past. Also, in this lifetime, after I got out of high school I had this great desire for having a home, the luxuries of life and would go to all lengths to obtain them, practically giving my soul away, and I feel I did help the negative forces in this plan to oppose the positive ones who were in the Confederation. We helped to overthrow Dalos. This really did damage to my psychic anatomy.

In receiving the readings of the lifetimes lived on this world, I can see the regeneration of the negative acts done. There is a pattern and the gist of it is, I was egotistical, proud, looking for glory. Everything was done for me. I was selfish, satisfying the self no matter how many lives were lost physically and their evolution stopped. I deeply regret deeds such as I have done that way and am striving to do better and be positive from now on. I thank the Brothers for their gracious help and especially to our beloved Uriel. (I know you are all in one but physically, you are here.)

P.S. The letter opener was similar to the knife used to cut the bodies used for fossil fuel and foodstuffs. I'm sure it was double-edged and very sharp. Correct me if not so.

About a month ago, coming out of the bedroom to the living room, this flashed through my mind: "Gee, what a life you are living, just like a robot!" I've pondered over this and came to the realization that I had a great part in making robots of thousands and thousands of people. It's mentioned in several places in the Dalos dissertation.

Jeannie: This is a testimonial of another involvement of mine during the Unholy Six. I was a scientist who experimented with humans. I believe that I used electronics and bionics to see what might be possible to create of a person. I tried to make robot-type people. I also tested the brain, trying to find a way to stimulate the regions not used by man.

At the Wednesday night meeting, Tom read part of a newspaper article in which a scientist had attempted a head transplant on a monkey, and concluded that someday this possibly would be a success. I felt that I had attempted similar experiments during the Unholy Six. I used people as guinea pigs for the 'benefit' of a negative science.

I also feel that I was involved in genetics, testing combinations in order to perfect a human for whatever purpose desired. In this present lifetime, I have also been interested in hybridizing plants; crossing plants to find a unique combination. I also have been interested in grafting plants and propagation, all clues to this reliving. I know this is just a start, but I feel more realizations will come easier now that I've begun to face this past.

Christine: Dear Uriel: When you corrected me

at the door of the Center when I said, "I feel like we're little girls," (Crystal, Jennifer and I, when you were treating us to lunch), I knew I had something to learn. That's a way my lower self has of wanting you to feel old, separated from us, inferior and left out. Here we are - the three fair maidens, etc. How silly!

I do really feel like an infant who has just found out that the future is infinite, but I never should feel inferior around you (you would never want me to) like I have in the past, a little girl. That's because I have wanted subconsciously for you to feel inferior around me! It is from the Dalos cycle when I wanted to make you think you were inferior - all those negative jabs. Today I was subconsciously jabbing at you in a very subtle way, trying to make you feel old. It was so subtle, I barely caught it. But Uriel, you are ageless to me; the most beautiful woman anywhere; exquisitely beautiful and beyond! So, my apologies. Thank you again for exposing another part of my lower self. Also, during the Valneza filming, I came up to you saying, "Your lipstick came off." You said, "Older ladies like myself don't wear bright colors on lips," etc. I had a reaction with that; it brought out my guilt but I didn't understand till now. I also have a fear of growing older which I am working on, finding out why.

Harold: Dear Uriel, I've read at least a dozen times the *Unholy Six* as revealed through Vaughn. Through much conceiving and analyzing as to my parts in connection to the harm done to Dalos, several realizations had come to me. Those realizations had caused me to feel that my main part was in the use of sound as a means of torturing Dalos. It was the mini-reading concerning my time in Atlantis as a scientist in the experimentation on human hearing that had brought about the rest of my realizations. Due to my hearing impairment, (nerve deafness), I can't hear sounds of a high pitch. So I feel this may be due to the use of the ultra-high frequency sound for torture.

One thing that really solidified this awareness is

my hearing test. The audiologist thinks that a hearing aid wouldn't help me because of my hypersensitivity to certain sounds. During the test, I had problems detecting low sounds because of my head noises. Another thing is that I often entertain myself by uttering funny noises while singing and I often annoy some people on purpose while doing it. Still more, many people have trouble hearing me because I often talk too low (part of my guilt complex).

Still another of my annoying habits is reading aloud as if I'm trying to tell someone something or to convince him. My voice would often get louder, as if I'm trying to emphasize something, yet I usually say no more than, "Hello, how are you?" or "Bye," etc., when passing by other people. So, it seems to me that in my past I have done some brainwashing and used propaganda.

Due to these realizations on my part toward Dalos, it really hurts when I'm aware that Uriel was that same person. I can truthfully say right now that this relating of such horrible times in the worlds of Orion, especially my destructive lives there, is the biggest bombshell. It really makes me wonder how Uriel and the Infinite could be so forgiving and let me exist. While writing this, I feel so terrible and tearful, it's so hard to express my feelings. I'm trying to remain objective and keep control of myself.

I would sincerely be very appreciative if the Brothers would confirm my reaction and relating and add more to it if necessary. Thank you.

Sometimes I'm lax in my study and practice. However, I'm still trying not to give up. Must keep on going.

(Note: This man is a bed-ridden paraplegic - writes with a pen or brush in teeth or toes, all due to these horrendous lives lived.)

* * *

Chapter 3

LIFE ON THE COMPUTER PLANET

5/30/79

Jeff: "Okay, I guess I'll get the ball rolling since I have a big, negative ball. Last night I decided I would sit down and try to write a description of at least one of my lifetimes in the Orion Confederation, and I came up with three lives. I got so 'into' it, it is a little freaky or frightening, really.

"My experience started yesterday when I stopped in a bookstore and started reading some science fiction books to see if I could get some ideas on art for the spaceship we are working on. On one page, a man had a picture of what looked like a castle. It was very science-fictionish and there was something about it that made me really sick. Across the front of it was written, 'Delirium'. Well, that tuned me in, and I am going to call this pleasure planet in Orion, Delirium, because it best describes what was going on in that planet. It is so totally unreal, it's as low as I can conceive at this particular moment. The whole planet was just one huge orgy.

"It was an orgy not only of sex, but of the most sadistic, vulgar, dehumanizing sorts of things that anybody could imagine. People were shipped in from other planets and if you happened to get your jollies from watching a person's head splatted against a wall, you simply sat there for a couple of hours and watched that very thing! Whatever could turn you into a complete animal, whatever were your wildest, most debased imaginations, they were brought into real life in front of you. People were just treated as part of your fantasy come true. People had operations to make them look bizarre so they could have sex in any kind of way imaginable. It seemed totally chaotic, but at the same time, I know there were people who operated this planet and whose job it was to manipulate these people who seemed to be going in totally random ways. They had to be controlled so the proper information could be gleaned from them - whatever kind of information. A person going totally berserk and allowing himself to become a seething id, makes it easy to find out where his true loyalties lie or to find out what he would do in a certain set of circumstances. The whole thing was precalculated to expose this person to the core, to know exactly what one was dealing with so the individual could be handled at a future date or could be eliminated if his loyalties were not one-hundred percent with you.

"Something I came up with, (since I find it necessary to soften this up a bit and in a different vein): Ever since I can remember, I will get this little voice-over-happening in my head, every once in awhile when I am 'down'. It is along the lines of the 9th inning in the World Series and he's up to bat. Well, I'll be doing something dumb like washing dishes in a restaurant and this voice starts telling me what a good job I am doing and that I am the leader of the world - the world's best dishwasher! I'll be getting into it, but I realized, a little while ago, that this was being hooked up to the computer; that the computer, no matter how menial your task was, continued to constantly stroke you. It was a stroke-machine that just kept you going and going and going, moving you in this direction, moved you in that direction, and I have still got part of that computer that strokes me for doing the most stupid tasks. Another example would be: I would sweep the Center or wash a window and I would expect to be stroked for it. I would expect someone to say, 'Jeff, that is the cleanest window I have ever seen - gosh it's a good window!' Baloney! It's a window; it's nothing, but it's something that is real in my psychic.

"I started writing in the third person, and then realized that I wasn't writing about 'them' but I was writing about myself and my psychic now. So, I'll get into it here: There were no personal relationships in this particular lifetime, that I tuned in to, of any kind. There were no families. People were born into a job. You were trained and looked after, to a point, but from day One, you were being trained with a certain job in mind. I was being trained with a certain job in mind, and I know I helped set up the program to make this planet work more efficiently - like a well-oiled machine. In at least one lifetime, I became a victim of it.

"There was no such thing as a home or a personal set of quarters that I could identify with in this past lifetime - it was total anonymity. I was kept moving and off balance at all times. The only constants were your work, your training and the Empire.

"While I was writing, I realized that this particular lifetime I was remembering was when I worked inside the computer moon we had operating. All physical wants were looked after, but in such a way as to manipulate the individual to do more. The computer was needed for things even as trivial as a glass of water or to open a door. The world consisted of building after building, after building that seemed to cover the entire planet. There was no sense of anything outside of the buildings. It was a total ego-expressing and anything natural or of nature - wild, so to speak - was looked upon as vulgar and disgusting. People strove ardently to overcome the natural looks (here I am getting into some parts outside of the computer), strove to look unnatural by shaving their heads; by painting themselves, decorating and dyeing, adding artificial parts to themselves, and generally presenting themselves as anything but what they were (human beings), nor did they have any spiritual aspirations that were valid. Our desire was to 'become at one' with the great computer; to become machines that were not flawed. As we were human and flawed, to become unfettered with seemingly irrational feelings or dreams of personal preferences was our goal. The worship of the machine was mandatory; it was logical, superior in intelligence and it served the Empire in a great way and therefore was a superior Being! In the same way that the Brahmans wish to become a sacred cow, so, too, did we of Orion wish to become a part of the computer machine. We are advanced enough to know that after death, in consciousness man was and is an energy wave form intelligence. This notion was perverted so that people literally prayed to become part of this computer bank after death! So of course, many did become just such an entity, obsessing the moon computer or incarnating as one who would only work within it.

"On this moon, as I remember it, my day from the moment I awakened was completely mapped out for me. I was fed my stimulant and only those in the positions of the highest responsibility were given any options or decisions to make nor did they necessarily start in the morning at sunrise or end at dusk. People went to sleep, thanks to a knock-out drug or a beam, at any location so that no one had any sense of day or night or location. The machine would direct me to a doorway, a hallway, an elevator, a rapid transit; then to another doorway and another hallway and into a room where my training was needed. This routine would be repeated as many as a dozen times, during any one waking period. In my travels I would see people likewise directed, but seldom was a word spoken - if ever. Certainly no word was spoken frivolously nor in any way of greeting. At times work needed speech but it was avoided; no one cared to talk. It was looked upon as eccentric or even ill and antisocial behavior - most unpleasant.

"At precisely calculated intervals, I would be directed to a dispenser where exactly properly balanced proteins, minerals and vitamins slid out to me in the smallest and blandest possible capsule or drink. It was something designed to speed digestion and also to stimulate me for the rest of my working day; to improve my efficiency. When the physical needed rest or there was no more work needed to be done, which, often as not, I had never seen before, I was again put to sleep with the drug. It would be either a few minutes or a few days before I would awaken. There were no distinguishing features to show the difference between one passage and another, one room or one building from another. Access was via underground tubes, mostly, with no markings around the doors or the various transit stops; only the computer in mind saying, 'Go left, go right, go straight; stop here, perform function A, B, or C.' The eye of the computer saw all that came and went on the planet. No one needed to know where they were going, what they were doing, because the computer knew . . . the computer knew.

"In that lifetime, I was a walking machine whose function it was to serve the greater machine. Some sleep rooms were for other things than for sleep. As I would sleep, I was given a complete physical examination or trained subliminally in some other facet of my job. Sometimes my mind would be explored for one reason or another. Once I was led to a room, put to sleep, and I never awakened. A special cot slid into the wall with me on it, and I was converted into the various nutrients needed in somebody's lunch. It was the day after I had fixed the monitor for a room that had panels and buttons that a man was manipulating. I could see him walking below the monitor, and for a moment I felt different. He looked different than the faceless bodies I had seen before. Suddenly, I knew what it was to be watched all the time, and I felt an empathy with another human being as being something other than a cog in the wheel. I felt superior to the machine - just for a moment. I felt the man was special because he was alive, even if he didn't know he was special. It was all very strange and peculiar, yet strangely familiar.

"For the rest of the day, I watched closely the faces of those in the transportation. I even missed my train once. I saw people as cups in an ocean and once the unity was realized, it was easy to flow out of my own cup and into someone else's. I knew their minds, felt their textures, and enjoyed immensely being a part of them. I wanted terribly to touch someone, actually physically to touch someone; it was wonderful and exciting. It was a longer period than usual, getting to a rest area that day, although I didn't really have any knowledge of what time it was; it seemed to take a long time.

There were no minutes or days; it just seemed to take long. That was when I lay down and went to sleep and never awakened. Some part of my psychic became a threat to the machine so I was eliminated. I began to want to form personal relationships. The really tragic part of this for me is that I know that I helped form that program. The moon had to function like a machine and it needed digits, and that was what I was, a walking set of digits. I would manipulate little dials and was trained for things."

Stephen: "That sounds like "Brave New World" by Aldous Huxley."

Jeff: "I'll tell you, I felt like such a robot when I finished writing this."

Hugh: "There have been so many things that have happened to me lately that I would take up too much time to go through them all, but I will try to hit a few of the high spots. This Severus reliving of the character that I played in for the television filming . . . I didn't play it strong enough; I didn't play it negatively enough because that was the character I was back then at one time in the Unholy Six. In realizing that, both in the rehearsal and in the play, I released a great deal of negation. And this was the person that other people were seeing in me and that I was hiding from myself.

"Well, after that I was able to get my readings for myself in a rather easy manner which I hadn't been able to do before that negation was removed. Next morning I would see things and look at them in a wholly different light and analyze them. One of the things that came to me was on Labor Day, when I was walking along the street wondering about economics; economics has been my forte in this present life; it was my best subject in school.

I know I had a lot to do with that. I suddenly realized that I was like the blind-folded donkey that had a harness on, with a carrot in front of him and I was trying to reach the carrot, and never having had the intelligence to back up and see that the carrot was controlled by me. I started going through all the economics of the world, and we all know of them. It did break through my money karma because yesterday quite a good thing happened to me in that respect. It will materialize in about a month but it was the ideal thing to happen to me.

"I started analyzing the economics of the world, where we are kept in a state of anxiety as far as money goes. There is plenty in the world, but we have to build cars that will self-destruct at 70,000 miles to keep people eating in Detroit and all the other places that supply Detroit. If you need $3,000.00 at the bank, they will give you $2,100.00, so you will have to sweat to pay it back. The whole principle of economics of the Earth is that there is not quite enough, to make it good. You have to really work at it, and it is set up that way by the people controlling it.

"I heard somebody say the other day that they were very surprised to find out that the Federal Reserve Board is not controlled by the government, and studying economics, I told them that it was true. The government has to ask the Federal Reserve Board if they will please do this and that. But they make a short supply to control people's overindulgences, which is a very reactionary system. I was analyzing that, looking at it and thinking about all the complaints that I had ever had about it, and realized that I helped create it. I had never looked at it this way before - that I had helped make this economic system that we have!

"This is how we controlled the world, because if we had given every person in the world that we were controlling, in the Unholy Six, an infinite supply (so to speak), and everything they needed, we wouldn't have been able to control them anymore. It is like we are controlled now, by the governments and the banks and everything else. But that is the technique we developed and which is in everybody's psychic. So that is where the economic system came from - the Unholy Six.

"Regarding the bust of Uriel I have been forming: It's drying out now and I have quit working on it - finally! So it's drying out, and hopefully, it will be fired successfully. But the workouts with that probably number into the thousands because I would have things flash across into my mind. It's like Vaughn said in his class last night, 'It's just a little fine line, going across; you could miss it if you didn't bring it back and look at it.' I would see the sphinx; I would see destroying or marring it - speaking for myself. There were many sphinxes built - small ones. I would see the cat people. Each time I would see one of these negative things, right after that I would see Uriel smiling at me. She was always there when I was working on her, always there and she would always have a twinkle in her eye. She was showing me these negative things I have done with sculpturing - even related to people.

"It has been said by the teachers here that we are doing now, what we were doing back then in the force field time. I thought, originally, I had something to do with the force field but what I really had to do with it came out the other day. I have had trouble drawing out the statue and I didn't know how to handle clay. Clay has been a big workout with me because I have used clay so negatively in the past. So therefore, the statue had to be made of that material to have my workouts brought to my awareness. But here I was, in the same building with this and having all the problems with it, and the dress for the Cosmic Generator was being made in the next room. So to me, that was like Dalos being in the next room in a force field, I being in the next room where I could see him.

"Several students came in when I was working on it and as I was having one final problem with it, which was the area across the bridge of the nose. Everytime it would start to dry, a crack would appear there. Well, I would try all sorts of techniques to fill the crack up. Everytime I would try to do it (and this must have happened 25 times), the crack would reappear. This is pointing out to me some pretty heavy karma in that respect. I saw many things. One thing was my doing plastic surgery, and I have been fascinated with plastic surgery in this present lifetime. In fact, I talked one day with a plastic surgeon and he explained the techniques to me and I was interested in it all.

"The replicas of Dalos were people. I told one student here that I felt I had made her look like Dalos. 'Yes,' she said,'Why did you have to cut me up all over? Why didn't you just get somebody that looked like him?' It then came to me that we couldn't get somebody to look like him unless they volunteered because we couldn't force somebody to mimic every nuance of a person's personality; not only personality but be suitable in outward appearance which others would see. During this realization time, I have a thing where I walk like other people. You just name somebody and I can walk like him. So it is like an artist who does caricatures and you exaggerate characteristics to bring them out - little things. I realized that when I was doing that with some students, Not only had I done plastic surgery to make people look like Dalos, but I had coached them on mimicking him to the smallest degree, noticing every little thing about him.

"So hopefully, that statue of Uriel will be fired properly. I have it the way I want it as far as I am able to do. And after that one is fired successfully, I want to do Dr. Norman, and after that, Tesla and Muriel. Then I wish to do other things that have occurred to me which will come in time.

"About the bridge of the nose: The plastic surgery was done there and one of the students pointed out that there was something across the bridge of the nose that was not done successfully. That tuned me in to soldier lives. I was asking myself why I was having this trouble with the bridge of the nose. Well, I was a soldier in many lifetimes and depending upon what weapon we had, I had a lot of pride and ego in using my weapons properly and calling my shots. Like in this lifetime, I am a marksman with a rifle, and in other lifetimes I have been skillful with a sword and this is where the break in the bridge of the nose came in. I would go into battle and delight in hitting people across the nose; not enough to kill them, but to blind them so they would become helpless, then I could just chop them up at my leisure.

"I had a reading on this foot-itching that had to do with my being the marksman who was selected to shoot Achilles in the heel. The shot knocked Achilles off his horse. He was such a good warrior, that was the only way he could be subdued. I relived this in many lifetimes and where I was also making swords and also making electronic devices for warfare.

I would always sell my so-called abilities to the highest bidder for whatever money I could get. I have even considered such in this present lifetime."

Christine: "You were talking about breeding women this morning. I didn't hear all about it."

Hugh: "It occurred to me this morning that women were bred as sex goddesses back then to grant favors to men, to get their cooperation on anything. These women's gene structures were studied and they were bred like you would breed plants or animals to the ultimate perfection of whatever was desired. Each society had a different feature; one society felt fat women were the pretty ones and some, the thin ones; whatever was desired was bred in."

Uriel: "I must say here that the statue Hugh sculpted is quite a likeness!"

David Keymas: "It is interesting you would bring that up, Hugh, because I have been trying to get involved in some college where I can learn how to work with hybrids, horticulture and botony. That was one word I kept on using - 'hybrid'. I don't want to work in a greenhouse; I want to work with hybrids and producing hybrids. I also got a flash while you were talking about the plastic surgery and mimic. I was always interested in the dissertation the Brothers gave and why it was stated that for two weeks Dalos was kept in the temple before anything was really brought to light (or dark, I should say). Before the force fields were switched on, for two weeks, many pictures were taken and I was curious to know why. There was some devious use for them and I realized that the people who would mimic Dalos and who would be studying him, would be photographed and these two video tapes could actually be played into a computer that had totally analyzed every physical movement, every structure, every facet of Dalos. It had analyzed that and it was stored in a memory bank. By syncing up these two tapes of the person mimicking Dalos and with the films of Dalos, analysis could be brought up as to how close it was and what areas needed correction. And this is something that just popped into my consciousness, that I feel very relative to.

"So I have had a growing realization of how we subdued planets without really coming into physical contact, just by starting a rumor on the planet. Last year, this was brought up, to a certain extent. We could actually start rumors, as it was stated in the dissertation, put lies in the government through the computer or other means, that another planet was preparing to war against a planet, causing planets to war against each other and nations to war against each other! In that confusion, we had no problem coming in and taking over. Or, after they had decimated themselves to a certain extent, we had that ground; one more planet, we had. We wouldn't lose any men; we wouldn't lose any ships. It was done in a very convenient and devious way. I have done this myself in this present life with other people. I would talk to one person and say, 'He doesn't like you and he has been talking about you.' That person would leave and another person would come up; the very person that we were talking about would come up and I would do the same thing about that other person who had just left. I don't do this anymore, thanks to the Brothers for making me aware of it. But this is something I have always been involved in and it is something that I wanted to rid myself of.

"The way I came into this is that I was accused of a crime at my work, of stealing some money, which I didn't do. I still felt very guilty with it. My boss didn't tell me anything about it; nobody told me anything. I just awoke one morning and the police called and said, 'Something has happened down where you work; we must talk to you.' They talked to me and told me they felt that I was suspect #1, and all the evidence pointed toward me.

"By what he said, I felt the people at the shop had told him things about me. This caused great resentment within me toward my boss. I was very much against them and very resentful. I felt they gossiped about me and told things about me that weren't true. So, I went in with this attitude of wanting to get revenge. Then I found out that they knew nothing about this. This is the way we would go about it in that past. A good presentation of this is in the Mang filming where various personages would appear on the video and say, 'I'm going to get my brother to boycott food for you.' We would have various duplicates appear on the video and say, 'We are preparing to war against you. Surrender, or be destroyed!' This wasn't the real leader of that planet or country. We caused a lot of confusion. I know this isn't all of it, but tomorrow I am going for a polygram test (a lie detector test), and hopefully, in this testing I will be able to get more on it.

"The police represent to me, the hierarchy of the Unholy Six, the bigwigs in the situation who would moderate these destructive acts through the power plays that would make people think that somebody else was doing them wrong.

"One other thing: I have never been able to relate to the Lens in a way of help. It has just been some abstraction that 'might' exist. I have never attempted to tune in to it for help or healing. Whenever I see it lately,

I would see it as a planet and I would see the Lens approaching in this manner: Here's the Lens approaching; here's the lens near a planet and this coherent beam would shoot out from it and the planet would be encompassed with this radiant, pulsating energy. The other night when this happened in my consciousness, the planet exploded. Now I know what this is! I am not seeing the Lens, I am seeing a configuration which is, as close as I can describe it, as the death star in "Star Wars". This is what I see when I try to see the Lens.

"So when I think of it shooting out a beam of energy, I am taken right back into this past and I am seeing a great war machine that was built. I sense that we actually had a great machine built that had, radiating around it, a great electromagnetic field, such that it appeared as a heavenly orb in the skies. It appeared as a star and it was called, "The Star of Good Fortune". It was supposed to bring good will to the planets but really, it was a gigantic weapon of destruction. We would go about to the various planets and say, 'We are here to bring you good fortune and we are here to help.' Within it, although it may have seemed beautiful on the outside, was a lot of destructive machinery. This has been the great block with me toward the Lens. I am not seeing the Lens; I am seeing my past. I am taken into my past everytime I see the Lens. Maybe I can wake up a little bit to the illusion of the past and to the reality of the future. Thank you."

<u>Thelma</u>: "As you all know, the next planet that we are going to visit in our filming program is the planet Valneza. Now I wondered when we first started, how a pleasant little planet could reveal so much negativity. Well, to my surprise, it is all there and it is now all coming out within me. There are a lot of

realizations I have had that I will not go into now, because I don't want to inhibit my workout when reenacting our parts in the filming of this planet. But there is one thing that has been very heavy on my mind. The first meeting we had, after we found out that the reason we worshiped the elements (air, sky, water, trees, etc.), is because we had controlled the weather. I immediately thought and said, 'Oh, this is where the sacrificing to appease the gods came from.' Well I kind of forgot about it for awhile but it kept coming back into my consciousness. I know I have had a lot of karma in other lifetimes with sacrificial appeasement to the gods.

"One day in particular the pain I had created in another lifetime where I was involved in sacrificial ceremonies came back really terrific, and I could hardly walk. I really do believe my maniacal, demonic self did demand human sacrifices at that time. Of course, the guilt grew greater, as in lifetimes that followed there were psychic memories of sacrificing humans to appease the gods. By this time, we had lost control of the elements, so it was all in vain. This has been really quite heavy with me and I really did believe that this happened."

Calvin: "I had a realization as we were working on this painting in preparation for the unveiling. In the Unholy Six Cycle and all during these times, I was involved in helping set people up, or make up individuals to look like gods or to look one way or another. This was done to make the other planets join us - planets that were for the Light and weren't in our confederation with the Emperor and who regarded our society as the wrong way to live because what our society deemed or what I deemed right, was the right way. This realization has come to me also because it is a very difficult thing for me, at times, when I am painting, to just walk away. And when I am in that state of consciousness I say, 'Calvin, just go for a walk and let it flow. Just allow what is going to happen, to happen.' This, as I have come to realize, is because in my mind I think I know what a Higher Being is or how an environment of that higher nature should be. The only reason I have those feelings is because I have these remembrances where I set myself up and helped to set up a society where certain individuals who were in power were looked to as gods. So, this is a realization that I have had about my own endeavors to turn around my past through the arts and why I have become involved in that field. Now, I am trying to turn it around."

Michael: "First thing I would like to say is that I would like to thank the Brothers for the way I feel right now. When I came through the front doors, I felt rotten. I don't feel quite as rotten now, thanks to the Brothers. My basic realization has to do with Uriel and my feelings towards her down through the ages, and where they started. All through my evolution, when I would come in contact with Uriel, be it in a family situation or just viewing her from across the street, she was always the one in the regalia and I was always the one left out. I have always said to myself, in fact, even when I first came into Unarius I had a very bad attitude toward Uriel. It was more of the kind of thing I have said through the ages, 'Well, why can't I be that? I know everything she knows; I know how to do that.' I don't think I have changed that thought pattern for thousands of years. The reason goes way back to the Unholy Six where I was, I feel, one of the impersonators or imposters of Dalos that were sent to carry on the appearance of the Mission of Dalos without the intent. Now I can agree with Hugh in that he was a plastic surgeon, because I had a great deal of interest in his sculpture of Uriel and how it was progressing, etc.

"As to how I was able to mimic every nuance and breath, etc., of Dalos, I was just illumined as to.

"Now most of you were here for the Lemurian Cycle and I wasn't. I read the books where it said that we had crystals implanted in our heads. Well, that's not the originating source of those crystals; they went way back, although perhaps they weren't crystals but the energy was the same from this Unholy Six time. In those two weeks that Dalos was in this fake building that we constructed to show what kind of people we were, and that we would benefit the governing body, what we were doing, actually, is what the Russians were just caught doing the other day at the American Embassy, and that was, bombarding them with high frequency radiation so that they were almost able to hear thoughts. We were able to pick up thoughts and since we had the photographic equipment, it was just like a big FBI story.

"Now that was placed in the computer, and since we had complete brain scanning abilities, we knew all the experiences Dalos had ever gone through. Through the crystal-electronic-type device that was implanted in my brain, and since we all were hooked up to this computer, any situation that I was in, my reactions were not my own but they were the ones that came out of the computer. But I do accept responsibility for them because I did volunteer for this position, because you couldn't make anyone do anything of this nature that he didn't want to do. So I volunteered for this position, thinking that I would get off the list for the gristmill, but I became just like a puppet on a string. The computer was telling me how to say 'Hi', if somebody said 'Hi' to me; how to say 'Hi' if I met somebody that I would have known at three years of age, such as Dalos. I would have felt that I knew that person. Now what this actually created within my mind, and to a degree my psychic anatomy, is the idea that I am as good as she, an erroneous idea. It's a negative vortex, but it's still there to be drawn from: I was that person, that Dalos replica, and that is the originating source for my feelings about Uriel."

Patricia: "I want to speak a little more on this monster subject. Well, I want to finally admit it: I am insane. Before I came into the Science, I had tried to commit suicide several times and I had nervous breakdowns, but I lied to myself and double-thought those actions out. Well, this Unholy Six Cycle has been such a great cycle that I've had to face this fact. I have had a few deflations lately and each time this monster that is I has taken control. It screams inside of me, it yells, and I almost want to pull my hair out. It is only these strands of Light holding it back from total destruction; from my spiritual destruction because if I were ever to let this person gain control again, it would be the end. It's not like I would do something like leaving for a day or a week, but if it were to gain control, I would actually commit suicide and it would be the end. But there has been great help in this cycle for me; so much so, that I have been able to resist this strong pull. With the help of Uriel, I have been able to put my mind upon these crystalline Jewels of Light and I know that any second that my mind isn't on the Brothers, I slip far away. In this cycle my mind has to be on the Light and the Brothers as much as possible. I know that after making it through this cycle, this Battle of Armageddon, that it will never be this difficult again.

"I also want to accept the responsibility for something that I haven't done in the physical, but I know that in the psychic, I have been helping to destroy that video that Dennis has been having problems with. I have said to myself, 'Oh, he is very negative to do that.' But I have had my consciousness toward this video, and I know that I am like an obsession, projecting out to him to do this. It's not just his fault, it's my fault too. I wanted to accept that responsibility; that thoughts are as destructive as actions, and I have expressed opposition. Thank you."

Edna: "I had a realization of Severus after we were doing the reliving and program; in fact, I guess it was during the filming. When we had the first act and we all were reliving fighting, I found it quite easy to let that negative energy force that I have, come through and I never realized before that it is actually an obsession from the past. I didn't know that those were obsessions - our past - and are actually our own obsessions! It wasn't until after the first act, when we went to do the second act and I walked in, that suddenly I was disoriented and didn't have that usual power or energy. I thought they changed what they were going to do and they didn't tell me because I didn't know when I was supposed to enter or what I was supposed to be doing there. I sort of spoke to Curley and said I was supposed to go in with him because I thought that was the original idea. But I realized that after I went in, I didn't say the words I wanted to say. It was as if I didn't have that energy or force. In other words, the obsessions seemed to have left for a moment and I didn't seem to have the usual personality, or whatever it is. I feel that this is going to help me a lot because I didn't realize that those were obsessions; that the energies we think are positive things that we are doing, and the energies that we let come in, are obsessions from our past.

"I feel tonight that as the channel spoke about kindergarten, that I am going to be in kindergarten for quite awhile. I have much studying to do to be able to face many things, and to not just guess but to let them be true. Something that came up when Mike was speaking about the crystal, sort of rang a bell.

"In the second act, Stephen decided I should have some ornament around my head and proceeded to find a trinket to put on my forehead, which also tuned me in to something from my past and I felt a bit robotized. Even when we went to the park to film and were watching the film, I thought I looked a bit like a robot, the way I moved. I'm sure it has been so in one of my lifetimes because I have realized there have been many lifetimes in the Unholy Six that I seem to relate to - everything; the astrology and everything. But I know there have been many lifetimes and I am sure I have much guilt. I am hoping with the help of Uriel and the Brothers that I am going to be able to face my guilts and know that they really are my guilts.

"I had a healing the other evening when they played the Moderator's tape. I was sort of in and out of sleeping. Then when they put on Uriel's tape of poetry, I started to cry and cried through the whole thing. I filled up and I know that was a healing from the Brothers trying to help me, because after that, I felt as though that force from my past was no longer present - the obsessions. I thank you."

Kathryn: "Ever since Patricia's testimonial I have started feeling very hot inside and shaky. My whole inside is pulsing right now, so I knew I had to get up and speak this.

"This is an experience I had with two of the students, Dennis and Jennifer. I always felt that Dennis and Jennifer's wedding should never have been. I felt that I should have married Dennis. This is the feeling and thoughts that would come into my mind often over this last year. About three weeks ago I mentioned to one of the students, these emotions that I had in me. Then two days after that I heard Jennifer say that it didn't matter if Dennis and she had a marriage license or not. I felt within myself that the thoughts that I had been having are true. So I was thinking about this, and I had come to the realization that within me, I was obsessing Dennis in this way; as Patricia was saying that thoughts are things and that thoughts are actually as powerful as actions. Well, thoughts are very powerful. I had this releasement and I let Dennis go within my own mind; psychically I have released him. I felt much freer. So this is my experience. Thank you."

Bill: "I am facing the fact that during the time I was involved in the Dalos reliving in the Unholy Six Wars, I enjoyed all the torturing she went through because it gave me a chance to go out and do my thing which is photography. At one time, when Uriel gave me the opportunity, she bent over to pick up something off the floor and the first thing I thought was I wanted to kick her. Then I went, 'Oh, my God!!! I thought that?' I felt that I would just be disintegrated at once. Then I realized that this is what Unarius is all about - to get these negative feelings out because this is really what happened. I feel that at one time, I was taking snapshots and she wasn't giving me the response that I wanted so I just hauled off and hit her or kicked her. That is why that came through so strong - for me to see that.

"Another realization I had the other night was: I saw this spaceship just tearing through space. Lately I have had . . . very subtly, I can feel this tremor. It is like Darth Vader says, about a 'tremor' in the Force. I feel it every now and then. I feel my physical body wiggle like I am going really fast through space in a spaceship. This is what I saw: going through space in one of these spacecraft, and the feeling I got was that I couldn't wait until there was another dogfight; I couldn't wait until there was another assignment so I would get to use this gun, this really neat, mounted gun that was a laser that we were using to fight. And the feeling was that it was not that I felt like I was hurting someone, but that I would get to use the laser gun; the obsession: to feel that power that it put out; to be in the position to blast it. So I have to face the fact that in another life in the Unholy Six, when I came back and used those techniques, that I learned in photography to zero in on something and catch it at the right moment. I used it in the weaponry. I have more, but that's it for now."

Christine: "I just now realized that something I did during the Unholy Six was to control other people on the planets or even people who were working for me, who were robotized. If we had control of their minds, we also could control their sleep states. Now, at times when I have really bad pressures from my past, I have horrible dreams that seem to make no sense at all. I know this is from what I did to other people. I was having a horrible dream the other morning and right in the middle of the most horrible part, I woke up. The tension was so great that I ground my teeth and I actually cracked my own tooth! So in the series of testimonials I have pieced together, one of the things I did was to make people's sleep state so miserable that they couldn't get out to the inner worlds; they couldn't get recharged and they couldn't think in a positive way at all. They soon feared to sleep and therefore I got more work out of them. I introduced nightmares into their sleep state and eventually they would not want to sleep for fear of what they would encounter. Sleep would terrorize the individual, having dreams of people chasing after him to kill him; monsters, etc. Then the person would become so pain stricken in his day-to-day life, he would have to go to seek some type of help, but there was no help, just a type of anesthetic. Insomnia victims in this present life could be reliving some such horrible past. I am working in a psychiatrist's office now which deals totally in blocking off people and giving them anesthetics for their pain or inability to sleep. So I know that this is what I did at that former lifetime."

Clark: "Hugh mentioned something about economics which rang something in me. In this lifetime I would read economics books and something would come out within me and say, 'Oh, wouldn't it be neat to make a perfect economic system?' This would be some type of a carefree system where there would be no problems. I discovered that there is an obsession within me to make perfect economic systems and perfect societies. So there is something that I really have to look at there. I believe I tried to influence the system of economics of a planet.

"Another area with me is religion. Religion is something I have a lot of karma with. I was also pretending to be a Higher Being to some people. So, that is something I have to look at in this period."

Dan: "One of the things I want to get out right away because I have such a big guilt with it is, that often I get Uriel water. At this time when she was Dalos and in other lifetimes, I would give her water that had human feces in it. I am sure it was my own. I had that realization after I heard sombody else's testimonial. The reason I know that is so is because sometimes when I wash her glass, sometimes I haved used the bathroom before. Then I would scrub my hands, and no matter how clean I tried to get my hands, I felt I was polluting her glass. So I am not going to do that anymore. That was a very strong guilt with me.

"This seems to be my week for guilt with Uriel. Last Sunday I got up and opened up a Pandora's box. Since then I have really let this demonic self come forth and I have relived much of the energy that I have recognized, that I talked about in my last testimonial. The Isis Cycle was a regeneration, a reliving for me of all these things that I had done during the Unholy Six Wars where I was this being who thinks he's a god and has destroyed planets and then allowed himself to come back and be destroyed because of his own guilt. This force came in the last couple of days, and especially yesterday. I was going through all these different emotions and realizations, and Uriel called up and said she was coming down. This gave my lower self a signal to put up a defense. When she came down I was totally reliving the energies of this old self. I couldn't do a thing about it; it was like I was caught up in this whirlpool of negative energy. Even though I knew what I was going through to some extent, there was nothing I could do about it; I was totally caught up in it. As soon as Uriel asked me what was wrong, it came right in. Finally I was so guilt-ridden and fearful, I didn't even want to be around her.

"Finally I decided I was going to sit down and face the music so I took a chair beside her at the table where she was talking to some students. In this lifetime I have had these realizations; in fact, it started last Sunday after the meeting when I went with Vaughn and Dotty to have something to eat. They had coffee and I had hot chocolate. It is just a little thing but the energy there was very definitely from the past - this past of Isis and Osiris. Vaughn got up to get the coffee pot and I went into an emotion of, 'I want this coffee and I want it free.' I had ordered hot chocolate and the waitress wasn't around and so I could get a free cup. And Vaughn asked me, 'Do you want a cup of coffee?' I said, 'Yes,' and he poured me a cup of coffee. Well, when we left I wanted to pay for the coffee. I said I had this coffee and I should pay for it, but I didn't. I was afraid to say anything to Vaughn or Dotty feeling, 'Well, they'll just think I am some kind of a kook wanting to pay for this.' I said I would just pay for it at some other time when I am here. But it bothered me as to why I had such guilt. Then it hit me that this was the energy of the Isis and Osiris lifetime when I had a desire for a position of power, and Seth subconsciously knew this. I wanted something for nothing and it was being offered to me so I took it. This was the attitude that caused me to enter into going against the Light.

"Now the reason I went against the Light, (I had a realization also), is due to factors that happened in the Unholy Six War when a certain person who was very powerful came to me, as in the script that we are writing for "The Arrival" movie, and offered me a chance to be a so-called god, or I could remain on my planet and be crushed. It occurred to me that I was going to lose my position if I didn't go along with this person. So this weighed heavily with me in making my decision, plus, I had this desire for power. I wanted to be a more powerful person, so in my own logic, I said, 'If I resist, he is going to destroy the planet and me anyway, so I am going along with him and be a god.' This is what I regenerated in the Osiris and Isis time. I had a certain position, as many of us did, as a priest in Isis' Temple, yet I had this fear that I was going to lose my position, that somehow I was going to be deflated or wasn't going to be able to live up to the standards of the temple and I would be ousted. So I was looking for another position. This is the memory of having this contact with this being in the Unholy Six, 'Come along with us or we will crush you anyway.' The act of having actually gone with him and going against that person after I had gained the spaceship and they had taught me to act like a god, precipitated running off on my own to play god without having to take orders from the people over me.

"I was killed; I was destroyed in that lifetime. Now I realize that from life to life, I have constantly had these energies of going against the people over me, such as it was with Zan in Lemuria (whose life I did live) a head of the cult, and interjected my own teachings and got pulled down. And this has been my repititious pattern. So I have this very strong subconscious fear of losing any position that I gain. So that has helped to release a great deal of guilt, fear and insecurity.

"Yesterday I was talking to Roberto (my roommate) who has been having problems finding a job. I have been interested and I have been trying to help him because I knew he was having a difficult time. I thought I was doing something positive and then yesterday I realized that I was reliving just bragging to show that I was superior to him. This morning it came to me that what I was doing is reliving the energies of secretly having a position under Seth in this past, and I knew that Roberto in the past was looking for a position. Because I knew that he wanted this position as Seth's agent and my own agent, I came to him and told him there was a secret plot going on, and why didn't he get in on it. That was the energy I was reliving with him; I know that's what I was doing.

"This is what I did: I went around as a spy in the temple, listening and trying to do the same thing I did in the Unholy Six War. I tried to find people who were dissatisfied and got them to come in and follow on my side which was the negative side. Another realization that I had is with these T.M. obsessions which came in on me again. I had this feeling that I was smarter than Uriel, all of a sudden, and I recognized that the frequency was from these T.M. obsessions. I began to analyze them, and as you know from some of my testimonials, I was one of the leaders of a T.M. cult, one of the top leaders. I realized that this started back in the Unholy Six, and these New Age groups like T.M., Scientology, etc., were started during the Unholy Six Wars as a way of infiltrating the planets, with seemingly something positive like this T.M.: 'We are going to give you rest and relaxation; you'll have blissful peace and we are a very positive group.' Underneath it all, are all of these obsessions just waiting for the poor sucker to come in there and they jump on him and hypnotize him subconsciously, and before he knows it, the poor soul is overwhelmed with these new philosophies and he doesn't know what is going on. Little by little we would infiltrate the different planets and take them over. I feel I was one of those in charge of this type of an operation."

Ronald: "At the last meeting Uriel's water glass was there on the bookcase with water in it, and when I looked at it, it seemed as if I could see little wiggles in the water. I felt like I was putting something in the water although I was just looking at it, but when I got up to speak, I had trouble speaking! After I sat down and analyzed it, I saw that this was a time of Isis when I drugged Isis so she would be late for the meeting, then I could get up before the group and try to influence them to follow me. These are some of the things I have become aware of doing!"

Brian: "Right along the lines of Danny's testimonial, just the other night I watched on television, something that has been on that I had seen listed in the guide that I had avoided. It didn't strike me that I was avoiding it due to something negative from my past. I just thought, 'Well, the show is negative so I will avoid it.' The name of it was "Psychic Follies". It was about many different psychic people and the various things that they were doing. I flipped on the television channel and started watching it. My first reaction was that I couldn't watch it. I said, 'Oh, God! I don't want to watch this.' Then it suddenly dawned on me that I didn't want to watch it because this is a good example of something that I should watch; so I sat there and watched the whole show.

"Even in studying Unarius, sometimes I lose the reality of other dimensions, or Spirit, of astral influences, psychic phenomena, and clairvoyance, etc. On this show were individuals and one fellow who had developed his clairvoyant sight to an extent, saw completely without his eyes open. He had his eyes completely covered and there was no way he could have seen, and he was walking around like he could see just normally and doing different things to show that this was a real thing that the guy had developed. There was a hypnotist on and another guy who implanted thoughts. He planted specific thoughts in a girl's mind to make a choice on a card, and of course she chose the one he had implanted into her mind. She thought it was her own choice and that she knew what card she was going to pick. That was just obvious to me. But the lady who hypnotized these people caused my strongest reaction because I realized the reality of the hypnosis. I was watching these people and was afraid that she was going to say something that would tune one of those people into a psychic shock of a past lifetime, maybe a fire or something.

"That person is going to start reliving that right there and she doesn't know what she is doing; she's starting a cycle for that person that could be really detrimental! The way she was flaunting herself around in an egotistical manner made it hard for me to sit through the show and watch it. It seemed so ridiculous for the people in the audience. They even had a real seance at the end; the table rose and the people had fearful looks on their faces because they didn't know what was going on."

Michael: "It was a fake. It's an old trick."

Brian: "Really? So, anyway, about the seance, knowing that some are real, tuned me in to the reality of it then - the people in the audience clapping . . . but on the other hand, one could do anything in the way of duping! The guy shooting these things was saying, 'Look how clairvoyant I am!', and they are all just little tricks. And this is the first time that it has hit me how fooled people are today. They were clapping like puppets. They had no idea what they were dealing with at all. They didn't know it wasn't real; they just watched it and still it was real to them and they were clapping about it thinking that it was really neat. It tuned me in to some lifetime when I knew something more than the average people. I was walking amongst them and feeling the superiority. They were being duped. We knew what was behind what we were doing and they didn't. So this showed me much along the lines of what Dan was talking about - instigating these mind practices amongst these planets.

"I know that I gained a huge amount of ego superiority through mastering these techniques of mind science and going to these planets and being able to implant thoughts and of being able to wreak havoc, and to have complete superiority through this mind science.

"Then sometimes when I am studying Unarius, a negation comes up with a transcendency that I might have. I'm sure that this is where it relates to, and even my getting into Unarius was really a motivation, as well as getting onto the true path, to regain this past power! I wanted to have the ultimate mind science and to have this power over the people, so it really meant a lot for me, and it meant that for once in this cycle, I can see the reality of people acting as sheep. It really made me feel sick to my stomach to imagine that. It finally dawned on me, the severe negation involved in robotizing people. Here's a soul who has the potential, and without him even being aware, you make a fool out of him more or less, because unbeknownst to himself, he is doing something so ridiculous in living these lives and all due to his selfish desire to have superiority. All these beings who could be expressing in a positive manner are doing something completely foolish in their development, when they really need to be learning and going ahead. So it was a good tune-in for me, to realize the reality of it and what it did to leave these people in that controlled state of mind! So this meant a lot to me, but it's just a beginning for me to go deeper into my specific involvement.

This was just an enlightenment to the reality of, 'Gee, this is for real! People in our society are like that.' It gives me an impetus to know what this is all about."

Christine: "This is important - for me to get this out. The testimonial Jeff gave about his realization where people would kill others just for the pleasure of it really brought home to me the truth of it. I have sort of pondered about this in my mind after something I tuned in to after watching the group of people killing Calvin who was acting as Severinus in our filming. They were up there and the group of people would come in and they would have to keep doing this shot over and over again. He would walk in and they would kill him and kill him again because Tom said they weren't doing it realistically enough. As I watched, I thoroughly tuned in to a lifetime where I was judging the killing of some poor person and it was all for the thrills of the audience. Then after I realized that, I realized that I had trained people to kill for the pleasure of other people who watched. I would sit there, watching as a trainer saying, 'Yes!' or 'No, you did it wrong that time. Let's have a little bit more of this gore and blood. Get him in the jugular vein faster next time!' It's total insanity but I'm sure these realizations are helping me!"

Patricia: "I have had a terrible obsession of lying to myself and believing my lies. Last night Vaughn asked for our diary of thoughts and I said I had done them, and I really believed that I did it. I looked the next day, to find that I hadn't done it. That brought home a couple of things: that I have been lying to myself and I believed all these things. What I tuned in to just now is that the computer could control certain thoughts. If you thought a thought or did something you weren't supposed to know about, they could just brainwash it out. That way you could be recycled. It was really a tremendous realization, finding out that I have been believing all these things that weren't really true. I couldn't believe my own thoughts!"

Vaughn: "Grand obsession. How many of us here have not worked on the Cosmic Flame configuration? If you would like to have an opportunity to do so, it will be finished soon, and it is going to be a very, very important demonstration and occasion. If not, you will feel left out. The reason is, you won't have discharged your own active negation with the Dalos Cycle - the incriminating evidence. We all want to work out completely, that cycle. So we have had a wonderful opportunity here to help you be free. Everyone looks so different and light. Let's keep it that way."

Chapter 4

Student Testimonials:

6/6/79 - Kathryn: I made a promise to myself a couple of days ago, and that was to Uriel. Within myself, I said that I would do anything for the lovely Archangel. The thing that I have noticed that I could do the most, for this lovely Soul, is to look at the past that she has presented to me. When she brings these cycles in, it is a great opportunity to look at these pasts. So this is what I am doing. She brought one up and it means a lot to me. It is very important that it did come up and surface itself. It has surfaced itself a few times in this lifetime in the area of gonorrhea and venereal diseases. I am very aware that this is a problem on this planet that runs rampant. We have clinics set up all over the place to have this checked. Well, in the Unholy Six Wars, we had an interplanetary sexual ring going and I assumed that I was involved because I know we were all leaders in the Unholy Six in one way or another. So this is one area in which I was very much involved.

We had a lot of electronics at this particular point and we used these electronics to make this big, round entrance to this particular room. The very appearance of this entrance was a huge vagina and people could step through this as it would expand and open so they could walk in. Inside, there were a lot of women. Also we used plastic surgery techniques. The women had their private parts enlarged - very much so. We would breed people from other planets that we would bring in and have in the forefront. The part that is so very degrading in this whole situation is: After these diseases started coming in, we had planets where we would put aside all these people who were very diseased. We didn't help them. The reason I can feel so much for this disease is because in this present lifetime I had some type of venereal disease that went through my whole body. It started one afternoon when my hands broke out with sores and my body started to swell; then, before the next morning, I was in tears, crying tremendously. I was going insane. The next day I couldn't walk; my joints were a size and a half and I was all swollen up. I was put in the hospital for seven days and had intravenous drip. This stopped the infection and it went away. So these people who had these diseases were in tremendous agony and pain. This is one of the places where it was incurred, and maybe before that we had these problems also, but it has helped me a lot to have these recognitions. Thank you very much.

Stanley: I would like to give a testimonial on the movie *Alien*. When this movie came out, I saw a poster which showed a picture of the egg cracking and it said, "In space, no one can hear you scream." I said, "No way am I going to see that movie." Then I heard from some other people who saw it and I got more intrigued with this creature that is in it. It is a biological monster or whatever you want to call it. I knew it was a horrible thing and I knew if I saw it, I would be scared and wouldn't be able to sleep for a few days or so. But I still had this big pull inside me to go and see it. All these people were saying how horrible it was and I was saying, "Gee, I really want to go see this movie." I had a realization, with the help of another student, that in the Unholy Six we created these anthropoid creatures; gigantic biological monsters. You couldn't even call them monsters; they were beyond that. The part I played in this is part of what I am reliving with the Cosmic Flame. We would create these biological monsters and then operate on them to implant these electrodes into their brain and different electronic devices in different parts of their bodies. This would be like bionics or whatever you want to call it. We had the ability, when we put these creatures down on the planets, to control them by turning a knob. We could make them very vicious or hungry, and not hungry, by turning a knob. We could make them go insane with intense pain and they would just destroy everything in sight.

This rings a bell of why I always liked (even though I thought they were the corniest movies I ever saw), these Japanese movies about the giant creatures - Godzilla and all this. There was one that shot out ray beams. I believe we incorporated laser beams with these creatures too, that could be activated either by us or the creature itself; I would think more by us. This is one of the things I believe I am reliving with the Cosmic Flame with these light bulbs. I knew the first thing they represented to me were laser beams but I didn't know how or where. With this cycle, it shows me how and it also shows me why I had this big reaction to the movie *The Andromeda Strain,* a movie about biological warfare. That's it!

Pat Gill: I was just given a copy of the dissertation that came from the Brothers the other day regarding the War of the Worlds. When I got to the part about the water torture, I really tuned in to that very strongly. I realized that I was one of those who thought up this horrible torture. It regenerated in the Spanish Revolution when I didn't do the torturing myself, but I dreamed up that dropping of water from great heights on the chakras of the body, which has given me my disintegrated disk. Now, I was in a good deal of pain the other day when I read this and was crying, and released it and my pain today has been practically nil. I am just hoping that it will continue to be so.

Also I have been away for quite some time from Unarius. When I originally contacted Unarius, quite a few years ago, I got a letter from Uriel saying, "Dear Unariun Friend, indeed you are and have been for a long time, though you have strayed many times." Well, this was one of my periods of straying and I allowed some negative force to come in my back door and keep me away. I think in the past I have allowed myself to be drawn away by some negative force, but in spite of all this - my not coming to the meetings - I studied on the side in secret, because I have been putting in many hours a day for all the months that I haven't been here studying, even to the point of irritating my father which really doesn't bother me, because this is the way I set it up. All of a sudden I realized this and the minute I realized what was happening, I was drawn back to the Center and here I am again, and it's good to be home!

Margie: I have never been up here before but I am going to try to read part of my testimonial that I have written on the Unholy Six cycle. It goes: Even before Unarius I have had thoughts on cannibalism and of the sex factories where individuals would go to certain places and receive constant stimulation. I have had thoughts of cannibalism where individuals were drugged and cooked alive, inside unbelievably large cooking areas. I must have worked in such places in the Orion System as well as being a participant in the negative sexual activity to enslave men. A couple of nights ago I had a horrible dream of seeing these horrible mutations, brown, rock-like creatures in the same form as humans. They were starved to half insanity, then left on certain planets (many hundreds of thousands of them at a time). They would chase after people and grab them and try to chew the flesh. The people would panic at such invasions. In the dream I was being attacked by a few. It seems as though they were very inefficient in consuming the flesh because of the way they were mutated; they didn't have teeth but they did have a mouth.

In relation to Dalos and the force field, I helped by projecting sound vibrations of negative intent at certain unexpected times that would cause Dalos great pain and anguish. The sounds and frequencies were monitored into the force field so that Dalos was prevented from sleeping. These frequencies also caused great pain in the fingers and bone structures. My job, as I have come to understand it, was programming computers, since where I work is on a computer register, programming individuals to work in factories, many hundreds at a time. During the hours of work I would find myself forgetful, discombobulated, an unthinking robot, extremely resentful of the people who would come through. When the children came through on the carts, my mind would really blitz then because of my guilt.

I have always looked at the economic system as programming people into the 'give and take' system, using commercialism and subliminal advertising, centralizing the middle-upper class, white American family as the epitome of society. I realized that this is just a continuation of the programming by the Empire to be a working, self-sustaining supporter of the Empire's motivations of universal world control. When the computer system blew up and the worlds were left without any free support, a 'give and take' system had to be developed; in a sense, back to the old drawing board.

Since there was plenty of everything, though it could not last very long, there was nothing one could do because it was so technologically advanced that if there was not a place in the planet to get what you wanted, there were two other things possible that they did and one was to rob and plunder other planets for resources, for there were plenty of people. People could produce people to live off other people.

Not very long ago I went to the Mexican border with Benno, and we realized that it was a reliving of going to the border of the Empire. After that, Benno asked me, if I had food to last six months, whom would I give it to, myself or one of the Mexicans trying to cross the border? Since the discussion of this turned to the moral fact of whether it was right or not to have these barriers set up between America and Mexico, I came to the realization that people were prevented from entering certain portions of the Empire where food was more plentiful that would last a certain length of time. Others beyond the barrier who were once supported by the computer planet Tyranator, were prevented from gaining entrance, and therefore were left starving.

I have been very sick with this cold for the past

four days which is, I think, due to the negation which is coming to the surface as to what sort of person I was then and that was an egomaniac. There have been certain instances in this lifetime when I have gone into an insane consciousness, not feeling that I am a god but the actual conviction of being in the position where no one could usurp my position. All those crazy science fiction stories of barbarians going from planet to planet, conquering, I can relate to, for it is these types of lives that I have lived. Electronics and technology became the means to the end - attainment of power and dominion. It was hopeless, as far as the people on certain planets were concerned, to prevent themselves from becoming slaves to the threat of absolute destruction - the high-powered, destructive forces of technology. You had to do what the Empire stated, or else! I know that I have little compassion or understanding for people in this lifetime. These things are buried very deep within myself because I have taken lives with no thought at all, with the mechanical press of a button. I know that I should not look at people as just things for the benefit of my own subversive, subtle manipulations. The system of always getting the other guy first before he gets you, became a way of downfall, besides the many other negative transpositions that took place. So I am very grateful now for the opportunity I have to overcome my lower negative self through the help of this Science.

David Reynolds: I have some more to give on the biological techniques in research and development in the Unholy Six cycle and the further regenerations in Lemuria. As a teenager I had a hobby of collecting insects or you would call it entomology - classifying and mounting them, especially butterflies of all sorts, and a variety of insects. I just realized today that I was reliving those energies of collecting different specimens, bringing them into the laboratories and finding out what kind of stimuli they would respond to, whether it would be different audio stimulations to enhance their aggressive tendencies or reactions to different types of drugs or electronic stimuli. So, last night I had this dream and in it I saw giant bugs - insects. I saw cockroaches about three inches high and about six inches long and six inches wide. It was very frightening looking. I was up on a concrete pedestal looking down on them and I was very fearful that they were going to get to me; apparently I escaped. The whole floor in this bunker-laboratory was covered with them and they were crawling all over one another. I have seen this in science fiction movies too. I saw a couple of them where the bugs would just ravage the community, and of course, they were carnivorous.

Also a scene that I saw that is not too pretty, a long time ago, concerns Steve Anderson and Benno's relationship. In Lemuria it was found out that Benno turned loose some giant spiders on Steve to get rid of him. Steve described this to me and I tuned in to it and saw these creatures or these specimens. It was a black widow spider.

To get off the subject for a minute, the most fascinating, favorite insects I collected were black widow spiders, in this present lifetime. I raised them in my room and studied them and their eggs; they had hundreds of tiny ones. Of course, I got rid of them fast!

I saw this black widow in a laboratory condition; the body of the black widow spider was about ten inches long. It was black and the skin of it was tough. I saw it being drugged and it was in a semi-drugged state, and I saw a person whom I thought was Benno, in that lifetime, pick it up by its body because it was so thick, and the legs hung down several inches and it became limp . . . in a semi-state of consciousness. He was showing it to some people with the attitude, "Look what I can turn on you. I can use this against you." The next thing I saw was Steve being overrun by these spiders. Steve was so terrified that he was in shock and running past equipment in a laboratory; on the sides of the wall there was equipment. He was turning over test tubes and all types of equipment, trying to get behind it to get away from these things. He was so dreadfully frightened of them it was incredible. I think I helped Benno plan that against Steve.

Christine: I feel guilty for getting up here so often but I feel that these realizations should go in the book, and I feel guilty for writing them in a letter to Uriel. So therefore, I am up here every week. Just from David's testimonial, I have to relate a past that I tuned in to in this present lifetime when I was in kindergarten. I had a playhouse in the back yard which took much of my daily time, playing dolls, etc., and one day up in the corner of this playhouse I saw a black widow spider. Immediately I knew what it was because it had a red hour glass on it. I wanted to destroy it but was terrified that if I moved it in any wrong way it would jump on me. Even though it was just a little insect, it looked bigger than life. So I went and got a broom and squashed it up into the corner and killed it. Well, that night I had vivid nightmares about spiders and awoke screaming. Even though my eyes were wide open, these spiders were literally crawling all over me. They were gigantic in size like Dave described. They must have been at least two feet across. These things were just huge and they were crawling all over me, biting at my face and devouring me alive.

Then the next night I had another dream where I tuned back to a lifetime where I saw myself walking through this jungle, and I was about 35 years old. I didn't look anything like I look now. I had three black ebony slaves walking behind me with cargo - my cargo. They were carrying it on top of their heads. The next thing I knew, there were savages all around us and they were cannibalistic savages. I knew then that my life was going to be over. And sure enough, what happened was the cannibal natives killed my three ebony slaves right on the spot. This was in the dream, but they saved me for later. They tied me down to the jungle floor and I thought, "Well, now what's going to happen? Am I going to be eaten by a tiger or lion?" That didn't happen but at night all these creepy, crawly, jungle creatures - creatures of the Earth - lizards, spiders and insects, whatever kind is in the jungle, came out and started eating me alive. The last thing I remembered were the opening and closing jaws of giant spiders, seemingly two and a half feet across, hovering over my face. And that is when I awoke from this dream. Well, from what David described, I feel this episode with these gigantic spiders had happened on the outskirts of Lemuria during an expedition but it was a regeneration of the Unholy Six mutations coming back to me.

But what I really got up here to say was, I had dreams in my past but I never knew what they were. One dream was when I was in high school. I was in my own house and the ground was shaking. I looked out the window and lo and behold, the houses were dropping off into the sky by sections, section by section, one house after another house, then another house, then another. The destruction was coming down the block closer and closer to our house. I lay on the floor trying to stay still with all the trembling and shaking that was going on - I was praying to God to help me. I thought the house would cave in on me and I would be killed before the house flew off into the ether. That has always been what has perplexed me. I had the dream a couple of times in my youth. Today I realized, when I had a very quick flashback of a past lifetime, that I was wreaking havoc on a planet. I was in a flying saucer, a spacecraft. We were shooting ray beams and cutting into the land, and we were actually lifting off homes or chunks of land, piece by piece and then degravitating the land up into the sky. The people would scream and scream and then be killed. I was doing this for fun, just to terrify the people and the population of the planet that we were attacking at that time.

Another realization, I had this morning when I was watching television. There is actually an organization here in America where people get together and play war games. They develop and invent war games to put on the market. They were all in this one big building having a marvelous time. To them they were in Seventh Heaven because they were able to let their fantasies run wild about killing, maiming and destroying people. Well, the interviewer came to one man and said, "What is it that you like about this? Is it that you get to be a general for a few hours and kill thousands of people?" And the man said, "Well, everybody has the desire to be God every once in awhile; they want to play God."

So I see that in one past lifetime I took part in developing war games for the people on these planets so that they could relieve their frustration about just being peons; in other words, they played these war games to pretend they were the 'big guys'. They had little men and little tanks and bombs and missles, and entire target areas in the form of games; they could maneuver these little things around and pretend they were gods. They could pretend they were the gods that we, ourselves, thought ourselves to be because we could do all this destruction. And so this is where the war games come from.

David Keymas: I never cease to be amazed by what happens when we have partial realizations and we come into testimonial sessions. What David and Christine said, answered a lot of questions for me. What Chris said about the anti-gravity, ties right in with an experience I had last night. I was listening to an Unholy Six tape that was given last cycle. Tom was channeling the Brothers and talking about the Immortals and our behavior patterns back then. I had listened to it a couple of other times and it came to this one part to which I always had reacted. It was stated that we had antigravity machines that would actually lift people off the surface of the planet into an airless grave. I shuddered all over, and I was half asleep. It is very hard for me to shudder when I am awake let alone when I am half out of my body. That really showed me that something was there; I really had a strong reaction. I have always had a reaction to dying and being left out in space, or maybe falling out of a capsule. That really helped me from what she stated.

About the spiders: I once saw a movie over at Decie's house. It was called *The Giant Spider Invasion*. It was comical in a way because of some of the special effects, but now I see how very realistic it is. Planet Earth was invaded by spiders, and the way they came down was in little balls that looked like meteorites. The Earth was infested with these things that came from outer space. They would incubate; they would crack open and a spider would come out. The spiders grew to an enormous size, even to the size of a house. That size didn't bother me so much because it looked so phoney and the little tarantula size didn't bother me, but when the lady opened up her drawer and one came out, the size of the drawer, that is when it really bothered me. This is the size that Dave depicted. I have always been fascinated with spiders too, and I have offered to get a tarantula for the Unholy Six part that we are doing for the *Roots* movie. Uriel stated that the crew members should have a black widow on their uniforms and maybe a tarantula or spider crawling around on the console. I said,, "I'll get it! I'll get it! I used to have a pet one." My friend brought me one, and I wasn't afraid of it. I would pick it up and it would crawl up my arm, and I would put it in my pocket and carry it to school. I would do everything but give it to the teacher. I can really see how we could infest a planet that way, and everything that has been happening ties into this. Just today I was reading in *Tempus Procedium* where the Moderator states how they have bombarded DNA molecules with radiation, gamma rays, etc., and have created mutants with fruit flies and frogs, etc., etc. This is the way we would have done it. I have had a strong reaction to that, thinking that I have done it somewhere. I have changed the chromosomatic structure. Michael stated in the last meeting, a workout that he was having with the DNA and his realizations with the DNA. Everything is so interlinked, so I know that we have created mutants and infested planets in the way that was depicted in this giant spider movie, infesting them with these creatures that were supposedly coming from outer space but were coming from our warships. So this helps me a lot in the reactions I have to bugs. Thank you.

Michael: With spiders and me, I have always been friends with them. I mean, I would say, "Don't kill that spider." I don't like to kill insects; I don't like to kill anything. Well, I always thought it was because I valued another little entity's life no matter if it was a bug or a person, but really I am just feeling guilty for having disturbed their evolution like I have done. I am sure that I was one of the people who made those spiders by going into the DNA. I am not sure if during the Unholy Six cycle we had a psychic anatomy viewer, but what I conceive is that we would get inside a cell and build the protein, block by protein block. The DNA in every cell in an organism is the same, so really, all you would need is one cell to determine what that organism is going to look like. I am sure we had a machine that through one cell we had fabricated, we could determine what the particular creature would look like, so we could do minor little operations and change it around and customize our monsters.

Jeff: A few nights ago I had a dream. It was after a week or so of really bad nights where I wasn't asleep and I wasn't awake. I was lying in bed and couldn't get to sleep and when I did, I couldn't wake up. I was sleeping too long and not feeling rested. I was really having a hard time of it. Finally one night I had a dream where Uriel came to me and said, "You know, all you are doing is just hanging around in a big astral cloud." In the dream she seemed to take a sponge that had myrrh or some kind of perfume on it and she rubbed it all over my body. Instantly, in my dream I was wide awake and could feel myself completely babbling away some testimonial about drugs. I woke up and couldn't remember what the testimonial was about, then I realized that I was reliving a time when I had been in an opium den. In an opium den you are not asleep and you are not awake; you are lying on your side with a pipe kind of hanging there. If you are awake, that means that it is time for someone to come and refill your pipe. That is all the life that exists; that is the sum and total of your life until your money runs out, then they kick you out and you work some more and save again so that you can get another few months in this opium den. Being unemployed as I am, it tuned me right back in to that whole thing of not doing anything and lounging around. I got right into it completely.

Well, the proof of the realization is that since then I have awakened every morning quite awake, much earlier with much less sleep, and I remember my dreams. I have realizations every morning that I write down about what I am doing on the inner. So that is a great proof to me.

Last Saturday I went to Disneyland with some people but I was very reticent about going. I thought there was so much work I could be doing and here I was, buzzing off to Disneyland. I felt like I was not doing the right thing but I thought, "I know there is a workout up there so I have to go up just for the workout." We were hosted, as it felt, by Steve Anderson who said he felt quite personal about Disneyland, and it was quite obvious that it was true. We went through practically every ride in the place in record time. It was unbelievable. I was thinking, "Here I am, a big-shot diplomat," and it was amazing because we got there just before the crowds and what usually takes people two days to do, we had done by noon! So I had this picture as we were riding the monorail, of Brezhnev or somebody like that being shown around in all these underground tunnels, and I started to get into it a bit. Then I thought, "Wait a minute, maybe I am a diplomat. What is this?" And when we went on the magic mountain ride and I was hanging onto two bars, it popped into my head, and I was beginning to get paranoid about Steve Anderson about this time. I thought, 'I am being monitored.

I am being put into a stressful situation here and they are seeing how I handle it. Why am I being monitored like this?" And it came to me: Now if someone was to land on Earth and was shown Disneyland and was asked, "Well, what do you think of Earth?" Well, it's not realistic, and that is what was happening to me. I was being shown around a planet or a portion of a planet and was asked to make an assessment of the Empire with that little bit that I had been shown. It had all the special effects, and engineered cities that were designed specifically to impress people and to make them fearful.

Well, it began to build and build, and boy, this thing was really coming alive when we got to the Bell Telephone building. It was incredible because there was this lady who talked like a very annoying robot while we were waiting. When we got inside, there was a lady who was less lifelike than the radio automatons that they have on the rides. This lady was just completely zombied. We were watching this 360° movie and there were troops marching, etc. Stanley said to me, "Boy, they are showing us all their troops." And I replied, "Oh no, they're not. I am not in that cycle anymore." I said, "Hey, wait a minute. I am in that cycle!" I think I was reliving a time when I had visited a planet, and against my own inner judgment I had gone, because once they had gotten me into their special environment and radiations, etc., I was a goner. I guess I went along with them because on the way home we were discussing how to make a better android. We were getting into discussing new hydraulic systems and how to imitate muscles using fluid instead of levers and things like that. We had it pretty much down to an art by the time we got home. Then the discussion evolved into what the psychic consequences are of an individual shifting his consciousness into a completely android body. We figured you couldn't do that because you need to have a homing frequency and the psychic repercussions would be monstrous. So the whole thing started out that I was being asked to join something, and then I was being impressed by the androids and I was helping to develop them. Then I was wondering what the karmic consequences are of something that I have done here. I know it must have been very impressive - this thing of making androids and trying to become immortal.

Anyway, when I passed over to the inner from that lifetime, I know I had a tremendous sense of guilt because I had been warned not to go; that if I went, I would be overpowered by these people. I knew they were no good before I left. What appealed to me was this idea of immortality; that was, I am sure, a thread that ran through everything that the Empire was doing. We all wanted to be immortal and omnipotent. It all regenerated into World War I where I became aware of the axis powers, of the tremendous negative force, and I took every opportunity to give a rallying call for people to get up and stop these forces immediately. I was really quite fanatical about it, I would say. These people had to get out and stop them. Well, this pressure at that time to get people to rally to the call was the fact that I had missed the chance for people to defend themselves and to become aware of this time of the Unholy Six cycles. So that is a great deal of pressure off me.

I would like to thank Uriel and the Brothers for the immense amount of help that I am getting every which way. It is just incredible. A few years ago, if I had heard about this, even my own testimonials, I would have said they were science fiction; I wouldn't have believed a word of them. But it is true; every word of it!

Mike: The last couple of days I have been feeling less like a student than I have felt since I have started. I was totally unaware until the other day when I walked into the Center and I didn't feel a part of it at all. The reason I haven't felt a part is because I haven't really been trying; I have totally been trying to avoid everything. When things began to occur that I should look at and analyze, I have avoided it and not even bothered to think about it. But basically, I think one of the big parts is biology. I saw the movie *Alien*. I had to get up and walk out because it was that bad. It was more than just a tune-in; it is a pretty negative movie all over. Basically what was impressed on my mind was that it was real. When I walked out of there, I knew that the creature they portrayed in that movie was not something made up out of the figments of their imagination. This thing existed! What I was trying to do in my mind was to rationalize it, saying it was just another innocent part of the Infinite; it's not really out to hurt anybody or anything. But what I was avoiding was: Maybe it was an innocent part of the Infinite, yet we took it and used it against civilizations and people. We tampered with it; it wasn't just a creature taken out of its environment but it was actually made to do what we wanted it to do.

Another part of what I feel is one of the largest parts of my karma is what I am doing right now - working on a spaceship for the movie. This spaceship is a negative ship with which Mal Var wanted to blow up a planet. What I feel I am actually doing is regenerating working on a negative ship that actually blew up planets, so in the small way of making the model, now I hope to be working out some of this so that I can turn around even more in the future. But for me, I think I had to turn around avoiding and running that I have been doing lately, so I feel this testimonial is important for me to get up and say what I have seen within myself. Thank you.

<u>Patricia:</u> I have been thinking about the mutants too, but I see something different. I see not a creature, but a virus. I see cells that regenerate at a tremendous rate of speed, that are parasitic and even actually corrosive when the person touches them and they can't be destroyed except within the 4th dimensional frequencies. I had a dream of this the other night where these cells were regenerating so fast and the only way they could be stopped was above, on the spaceship. We said, "We can stop this anytime. Just hand over your planet." But it was not just the

3rd dimensional frequencies; it was the 4th dimensional frequencies.

I did want to relate also something about the Isis and Osiris cycle. I have been going through tremendous mental pressure, and I was outpicturing it. I finally realized that I had to use the mirror principle and realize that this was coming from me. I was a member of the counter-culture, against Isis and Osiris at this time in Egypt. Although I wasn't a head, I was somewhat of an authority in this counter-culture which was rising up to fight against the Light. As for those people who were under me, I tortured them mentally and psychically in ways such that if they didn't come up with some piece of information, or they didn't complete some task, then they were made to leave or were tortured. This came about when I was so afraid that I wasn't going to be able to make it to the Valneza meeting and I would be told to leave; that I wasn't wanted and I wouldn't be a part anymore. I realized that this is what I have done to people, and because their mentalities were not very stable in the first place, they would feel such anguish and insanity that would intensely regenerate inside of them. This is the reason I have been feeling the mental anguish because I have caused other people to feel mental anguish.

Dan: This will be a great one to follow Patty's. I have been going through quite a bit of mental anguish myself for the last days, ever since I became aware that I have to find a job. I have tuned in to the energies when I left the Light and have suffered. I have had a lot of conflicting emotions. Basically, I have put myself in positions where I would suffer because of my great guilt for having destroyed planets and people's lives; therefore, as sort of a subconscious punishment or beating myself, I have sought to acquire jobs or positions that, in my mind were menial as a form of punishment. At the same time I would seek these jobs, I would feel that the people who were going to hire me were looking down on me and humiliating me. This was, of course, my own consciousness that

I projected out when I had placed myself above other people and had humiliated and looked down on them; therefore, I was expecting to get it back but I was only getting back what I had put out. The last time I gave a testimonial, I said that I felt I had to get a job as a waiter. Well, this was that frequency of punishing myself, feeling that I am limited and this is the only type of job I can get because my guilt was so strong. But as soon as I sat down, I realized that Uriel doesn't care what type of job I get; in fact she told me on the phone the same day that I was talking about it before, that this is one of the problems with people - they limit themselves, thinking they can only do one or two things; that there are an infinite number of things that we can do and express to work out karma and serve the Infinite. So I have gone through both extremes of feeling that I have to do just this one thing out of guilt and then also feeling that if I don't do this one thing, that I am not a spiritual person and I don't have humility. Therefore, I will go and do this one thing that I hate just to prove that I am not above everyone else; that I don't think I am above everyone else, but subconsciously, I really do; these things are in my subconscious.

I have been trying in a way to deny my own feelings and emotions that I have created from the past, so I have gone through these emotions in the last couple of days. It has been up and down, up and down. Subconsciously, I don't want a job and I don't want to work and so I have decided that I will try to go out each day and put in an application in at least one place. But I won't do it if I am negative; until I have come to grips with myself and realize what energies I am putting out and what I am reliving, it won't do any good to go out and look for a job if I go out with subconscious frequencies of, "I don't want this job and don't hire me; I feel inferior here," and whatever. So it has been up and down but today I feel pretty good and on top of it. I know that I am just getting back what I put out and created.

Without this Science I would really be in a bad position right now; in fact, I can say the energies I have been reliving since I have started job-hunting are the same energies I came to Unarius on. As those of you who were here know, my first job was washing dishes. For a year and a half I went through this feeling of inadequacy. I would carry a tray of dishes and I would feel like everybody in there was looking at me, "You poor thing." I would have tears in my eyes, feeling so humiliated. So it is good to know that I am not limited and I can do anything. It is just the way that I conceive these things from my past that make me so fearful. It is not anybody but myself that I am reflecting out and getting back.

Crystal: I just wanted to say that I am truly experiencing the principle of what we have done to others, comes back to us. I slapped myself in the face several times today because of my paranoia of what I feel as someone else is feeling towards me. I have felt and projected these acts and deeds onto that person and people. We are going to be filming a flashback tomorrow night of this cycle where I am Dera, which is really going to be an opportunity to change around a lot of karma. I think it is very interesting that the people who wanted to be in this program are my sisters - and I hope that we really become sisters. It will take time. I know that my karma with other women is very heavy because I have placed myself above them. I have always moved myself into a high position. And the part that I am playing from my past in this flashback is a daughter of a very, very powerful man during the Unholy Six War. I was his daughter and I had everything. I had my own spaceship that was built especially for me and it was a pleasure ship. It was a large city that went through space and it might sound like fantasy but it was a place where all the high V.I.P.'s would come for their rest and relaxation; essentially like they do in the Army and Airforce and all. Spaceship captains would come and we would have euphoric conditions, beautiful female robots that would satisfy them in any way they wished so they could fulfill any fantasy they wished.

Now I see myself as the big one; you know, this was my ship, something that my father had me working on as a representative for him. I was used because of my love for what I did. And this is the thing: I was actually loving it! Tomorrow night I am going to bring out these aspects of my character and I am going to let this entity be seen. I have been nurturing a very destructive and sadistic force. A good word for it is a bitch. I have these frequencies now that I am really struggling to be free of, but you can't struggle; you just have to accept. I know that I project these energies out still because it is I and my psychic energies that I have built for myself and regenerated many lifetimes. But it takes time, and I become impatient with myself, wishing that we didn't all have to dislike each other so much when these cycles come in. I would rather just have everything be euphoric and not face it. But it is very painful to know that somebody doesn't like you and resents you and hates you, but I have to accept that they are hating something that I really am. What they are seeing in me and reacting to is truly there. Sure, it is within that individual too, but the hard part about it is accepting it. But I am, and I hope tomorrow night will really clear the heavy air because we have been heavy in it.

Christine: I have had a lifetime and perhaps a few lifetimes, where I was worshiping the sun, or I stared at the sun and consequently went blind because I wanted to go blind, because I wanted to punish myself for something that I did during the Unholy Six Wars to entire planets. I have always reacted to news broadcasting where they will show an eclipse of the sun, because the audience they show on the news will clap and scream and they act like little children, as if some god is causing it to happen. But we all know from the Science that it is a mere happenstance from hysteresis. Well, in the past we had the technology to tamper with the electromagnetic fields of a planet and the belts that keep it stationary in its orbit through space. The EMF fields conduct heat and light from the sun to the planet. By interjecting frequencies that would offset the hysteresis from the sun to the planet, we would create total darkness on that planet so their crops would not grow and starve the entire population until they would accede to our wishes and come over to the side of the Empire. So this is one realization I have had in which I have been involved.

Gordon Jenkins: Jeff gave a testimonial about the Pleasure Planet in the session the week before. I felt very strongly that I was there but I didn't relate to the gory part of the Pleasure Planet. I related more to what was referred to as the sexual aspects of the planet. But that was just it, until one morning I was sitting on my bed when all of a sudden there was this scent or fragrance that hit my senses. As soon as I smelled it, all of a sudden, a big laboratory flashed to my mind and scientists were operating on a person, implanting these implants into the brain on certain glands there. It was a process of what you might call electrochemistry. It caused electrical stimulus and impulses to go into the brain when it was activated by the person himself. When it did this, the chemicals that were in this gland changed so that it excreted the chemical into the bloodstream, and then came out of the pores in perspiration and through the evaporation in the air or in actually touching a person as it went through his nostrils and he would breathe it in. Or if you touched them on the side of the face or wherever, it would go into their skin. With the opposite sex, it would be opposite chemistry; it would be as a love potion that the cosmetic companies are always saying about this perfume or that perfume, "They'll go wild about you," and so forth. Down through the ages the gypsies have had a strong belief in this love potion. Back there it wasn't something that you put on like perfume but it was done with the technology by changing the actual chemistry of the body to put out this fragrance that would make people fall at your feet.

Testimonial by Patricia Joiner in story form:
The advertisement said, "Journey to Space Mountain", but for me it was a journey into the past.

"Starfighters! Load into missile-craft on Stardeck Number 1," a soothing voice permeated the great battlestar, as young starfighters scrambled from every port of the ship toward the main stardeck. Denura looked down upon the half-android humanoids from the control tower above. "As you see," she spoke to the young men beside her, "We have the most technological battlestar of the sixth fleet. Each humanoid on our battlestar is directly linked to at least four separate sections of the main computer on the planet Tyranator. Some can perform up to ten different functions with little or no manipulation for us here at Control Base Number 1. With the new advancements of the Emperor's Computer Planet, almost every function of the ship is computerized." Denura pushed a button and the voice resounded again, "Starfighters released to resume duties."

"Tell me, are your Starfighters completely android or human?" Jargo the head of the inspection committee inquired.

"Physically they are completely human. We find that they have more agility that way but technically they are android for each one has an electronic componant implanted at the base of the brain to insure proper computer contact. Mentally, they have no control; they are but the computer's brain in a humanoid body."

"So they are not really human at all," he observed.

Denura turned and pushed another button, "Shall we proceed and board one of the missile crafts?" she more stated than asked, and walked out the sliding metal doors. Denura was tall and lean, for she, like those of her position, had special treatments for the physical appearance of the humanoid. She had all the attributes that a man could wish for physically, but not a man in the fleet would touch her. She had made her way up through the ranks with calculating coldness. Not unlike many of the others in command, she left behind her a trail of anguish, despair and blood.

"Aware of the great pressure, Denura forced herself to think of other things besides the visitors. This was no ordinary inspection crew. Could the computer possibly have picked up her fleeting thoughts? "Power, glory, rebellion . . . No! No, how could it be?" she thought in confusion. "I do not have an electronic implant . . . or do I?" Her perspiration made her feel cold and clammy as she entered the chemist's cubicle. Alone, she picked up a small vial labeled, "PHYXEE". She shuddered. She had used such a deadly weapon before, and now it might be necessary again. Her mind wandered.

"Raised from birth as a chemical engineer, she knew the deadly effects of this poison. For so many years she had kept this formula secret in fear of the spirits that haunted her, that part of her that she had no control over; that part of her which could kill. When she first discovered this drug, she was having a quarrel with a competing classmate. They had been working on this combination of molecular poisons for quite some time and now were near success. Then it descended upon her, an overpowering obsession filling her consciousness.

"Jonu," amazingly her voice was calm and natural, "Our success has come! Soon we shall be working in the House of Hathor with the greatest minds of chemistry." He just sneered. "Please," she looked at him with a seductive look, gently putting her fingers over his, "We have not been on the best of terms of late, but do come to my cubicle for a drink to our success." Perhaps there was something hypnotic in her eyes, perhaps he felt triumph over the enemy for he secured the drug under lock and key and escorted her out the door into the open hallway. When she estimated that they were about halfway down the corridor, away from the cubicle they had just left, she turned to him in feigned anxiety. "Jonu, I have forgotten my term papers, which I must present to the council in the morning. Please wait here while I get them. I will be right back."

"Once inside the cubicle she went quickly to work

finding the small vial and pouring a small amount of the drug into another small vial, then putting the majority back under lock and key. Placing it in the inside pocket of her lab coat she walked out to meet Jonu. She then came back to life on Battlestar D4-73. Even now she could still see herself standing triumphantly over his dead body. The next few days she went about her duties under the close scrutiny of Jargo and his crew. She could only bear the pressure with the help of electrotherapy (something like our shock treatments) given with the help of Condrel, the flight coordinator. It was a comparatively simple process, but she had taken to Condrel's company - quite a strange thing for a woman who was so wary. The process consisted of taking a small pill which would chemically stimulate the brain, then needle-like electrodes were placed on the scalp and treatment applied through electric current.

"Carelessly, Denura remarked, 'Condrel, don't you think the inspection crew has been here far too long for a normal visit?'"

"Perhaps they suspect something." His usually casual voice had a hint of sarcasm in it.

"Condrel, what do you mean?" she blurted out.

"What do I mean?" she then asked herself as she could feel no control over her emotions. She could feel herself slipping . . . she could feel something take over . . . she could . . . 'Condrel, what is happening?' Her stomach was turning but she couldn't put two and two together. The cold metal door slid silently open as three dark figures entered. She struggled to speak. 'Jargo, what is the meaning of this?' Her voice trailed off into the distance. His handsome face looked distorted to her; everything was beginning to look distorted. He held up the treasured vial.

"We now have the formula to manufacture - PHYXEE. We know that you have not been true to your Emperor, for you have held back valuable information. We need to know the effects of such a drug, and who better to test it on but a traitor?"

"Denura's confused mind dropped in consciousness, lower than she had ever known. She knew how painful the drug could be. It caused instant addiction, loss of balance, disfigurement and eventual death. Her only hope was that they would give her a large enough dose to kill her on the spot as she had done to Jonu so long ago. Such kindness was not known in the Empire. She knew she would become an example, a side-show freak on one of the planets to warn those who would dare turn from the side of the Emperor.

Patricia: Now I know why I reacted so violently to the trumpeteers and what I called the "Zoo-like entertainment" the night when the Cosmic Flame was first presented. I was remembering all these deeds of making a spectacle out of people, and here it is, come back to me. Right now I feel as though I have just been poisoned, for I feel in this long-ago past that I was the one I call Denura. I apologize if the story form is not apropos to what is needed, but this is the way that I can relate to this horrible past. I especially relate as if it is happening to me, so I know that I must turn this around and realize that I have specialized in making people into freaks, especially in using this sort of drug as I have and feelings toward others, although fleeting, that I wanted them to look disfigured, like half animals. I know this was done as punishment, and as a warning.

Letters From Students

6/25/79 - Dear Uriel: The Sunday Meeting of the 24th was most certainly the "Night of Nights!" Everyone was saying, "There isn't anything She wouldn't do for us! Isn't this wonderful?" Many times these words were expressed by our young students, and most young people aren't able to understand or appreciate what was given us with your exemplification of the Generator - the Most High Spiritual Generator which is what you, Uriel, are. But our young Unariuns *know*!

Anyone having difficulty picturing the working of this wonderful, positive energy dispenser should have no further difficulty; the expression was most graphic. This extra effort to make the existence of these most high spiritual energies reality, gives us hope (it gave me hope) for an eternity to be with more and more lessons learned and more ways to work out the negative past we have accumulated up to the present.

It presented me with hopes for a future still permeated with the help of Uriel, Michiel, Raphiel and Muriel and the whole Unarius Brotherhood. I know they will always be with me and all Unariuns to help us in our future to become better representatives of the Infinite. Gratefully, Helen.

6/26/79 - Dear Uriel: It was during the early 1970's that I suddenly turned into a burning torch as I sat on the sofa at a Unarius meeting in the home of Uriel. It frightened me terribly, it was so horrible, this reliving of a viewing or participating in burning a fellowman.

For the perpetrator of such a crime, or viewer of such a crime there is total karma, especially when he feels satisfaction in the scene and helps make it happen. This is again evident as I face my feelings of worry, fear and discomfort as I watched Uriel within the force field (the dress designed to emulate the great vortex of the Generator of Creation and Healing which She truly is). I feared she might be overcome with the heat of the electronic lighting or that a short might occur in the wiring setting the dress afire. This is how I attuned to helping eliminate Uriel in the past and I believe it was that she was burned at the stake, I helping set the fire in one way or another - perhaps lighting the fire.

With this realizing of my great guilt, the feelings within the very center of my psychic indicate the greatest upheaval I have ever been aware of, psychically! I can feel the gradual sluffing off of negation and a lightness and upliftment as I see Her Galaxy of Healing within my psychic oscillating in its golden emanations. The very center of my being is becoming realigned and healed. I feel new, basic life-strength, new to my awareness and it has been actively going on for the past hour and more.

I feel I have finally gotten to the bottom of my feelings of dislike for hot weather, often becoming ill from too much heat and as Uriel advised me many times in the past when I felt overwhelmed with hot weather, it was because I am guilty of helping eliminate by fire, the physical of the Great One who has come to us so many times to heal and teach the progressive way. I believe it was done with the motivation for power and self-interest for material gains. That my mentor and helper is the very one against whom I perpetrated such a crime, in itself is the indisputable proof of Her true nature - the greatest Messenger ever to walk the Earth! To say I am grateful is understating, indeed! - Helen

7/1/79 - Dear Uriel: The first thoughts that came into my mind when I knew we were in the Dalos cycle and that Dalos had been kept captive in a man-made force field were: In order to attempt to influence him to come down to the earth level he had been kept extremely uncomfortable in ways amounting to torture. I felt the truth of this realization and also felt that added to the planned manipulation were the unintended malfunctions of the force field which caused even more extreme difficulty for Dalos. Parts of the electronic instrumentation were to wear out from time to time, and the so-called human error that most certainly took place, I felt wrought even more havoc to that great, gentle Soul. Now, after hearing the last few testimonials regarding his possible final demise, I feel that his death was caused by malfunction by the operator or that there was error in the parts replacement causing Dalos to be electrocuted.

I wrote of my reference to the time I became a human torch and I believe I caused this malfunction or error in the manipulation of Dalos causing his death. I keep feeling that it was error that anything should be the cause of his death because why should his tormentors kill the goose that lays the golden eggs, for they felt that their manipulation of him was aiding their cause? Yes, I feel that I was the one who caused this great One to endure this horrible death and that I have been suffering the guilt and physical heat from that psychic malformation down through the ages. I would feel that I have always, since that time, been sensitive to extreme heat temperatures which I have been in presently. - Helen

1/9/79 - Dear One, the testimonials Wednesday were wonderful - Thelmas's about burned victims. When I spoke to Azhure and Margie, I started choking. Something grabbed my throat and I couldn't breathe. Azhure saw a group of flying ships searing, burning a large planet with frozen beams. Men, women and children were burned as well as the land. I was the leader; I pushed the first button. I was smiling and enjoying it. The people didn't want to join the Emperor's League. My stomach was sore and my left eye was very painful. Even my eyebrow was jumping and twitching. I was sick all day but feel a little better now. - Irene Brown

1/9/79 - Dear Archangel of the Universe, I've read and reread the latest Unarius publication concerning the Unholy Six star systems of Orion and it is staggering and horrible; almost beyond comprehension for a mind to conceive - and yes, I was one of the perpetrators!

I've always been sensitive to changes in temperature

especially these days, involving my right side. Therefore, I may have been involved among the culprits who cooled the air molecules, causing freezing sensations. Then rays of great heat would be brought in causing great agitation to Dalos, who is Uriel this present lifetime. This was done alternately from one day to the next.

Yes, I feel I was one of those of Orion who knew of all these plans and secrets to torture the great Ambassador, Dalos, a beautiful speciman of a higher Being. We lied and spread distorted concepts that he was disrupting truth and that the Orion culprits were the innocent people, when the reverse was truth instead.

I realize now that a great plan, a purpose in life must be completed or static energy will be the result, and I realize since reading this latest paper, that I set in motion a plan to help Unarius with funds which hasn't been completed, causing me great frustration and guilt - hence all the depression, heart tremors, etc., that I've suffered from.

In the rest homes, two in which I spent some five months, I saw many people in varying, decaying states of both mind and body. No doubt, many or all (?) of these were my former victims during that horrible Orion cycle.

Although I have been told that I wasn't a scientist in any civilization, I've been a very negative individual, no doubt willingly doing the bidding of others for gain. I may have advised and schemed to distort the truth when I knew better.

This is my confession and I hope that it will help the thousands of people I've previously helped to lead down the garden path. And I do ask the Brothers to help me as I struggle to keep my sanity in this insane world. I need inner strength to accept and dismiss my horrible past. I hereby dedicate myself to the Infinite Truth. I am sleeping much better than I have for the past two years. - Ruth Trager

---

Chapter 5

The Great and Extensive Means to Which The People of the Orion Empire Went to Try To Make of the Super Celestial Being, Dalos, a Humanoid!

The *Cosmic Flame* Becomes a Symbol of the Vast Reconstruction Program of the Empire

6/27/79

Arthur: "Greetings and Welcome to the Unarius Center, the New World Teaching Center! Usually we have one of the subchannels moderate the meeting. He may speak later about some of these concepts on philosophy and what we are really involved in here. There is a recognition of our personal identity in terms of our continuity from other past lifetimes. What we are really involved in here, is what all people are involved in who are not yet at the stage of acceptance, but we are an example in our own dedication and in our recognition.

"We are validating these principles of reincarnation, the continuity of life into Higher Worlds, higher states of consciousness, new dimensions of awareness and truly, of more intelligence. These identities and realizations are now being filmed on video tape as well as on 16mm films. There has come through - from the very advanced, intelligent people who are the byline of our world in terms of teachers, physicians, scientists, writers and those we look up to as being humanists who have aided us - a semblance of a new awareness of the nature of evolution. We know who these people are by the names of Socrates and Plato; we know of them in terms of the philosophers; Herodotus is one of the historians. Among the scientists are Einstein and all the various sages have come through also who have proven from the basis of their own personal research into self that life is a continuous statement, and it is formulated each time in a new setting for one purpose alone.

"That purpose is: to be engaged in the true essence of the nature of our own Inner Self which is the Infinite Creation, the true energy definition of life. So it's an entirely new orientation for those who have not yet come to the point of accepting the dimensional qualities of life. They live in a certain consciousness, based on the 3rd dimensional principles of physics but we know that this is only one of the planes or dimensions. There are an infinite number of planes and our life, lived in a positive, progressive fashion, will be based upon an awareness as to the function, nature, and purpose of this great energy flow which is generally called the Infinite Creative Intelligence.

"Specifically, it is the consciousness that we have at the present time which gives us integration with our present life here and now. So this is the work that is set up for all people: to learn their true identity by engaging in life, and through the work that is involved, we here are basically the pioneers, helped by these great advanced, intelligent Beings who have set the pace, who have provided the means and have actually given us the textural framework so that we have a methodology, a technique, and an operational way of studying ourselves. Now, because of this and because it takes a certain degree of orientation to conceive new principles, and since it takes a development, it may shock certain people who have not yet had the basic or the mental preconditioning as to the continuity of life; of the framework of life as being vast, beyond the present physical relationship of life and what we call our basic life patterns from one day to the next.

"Uriel, our Spiritual Teacher, suggests that those who wish to come to these twice-weekly meetings have a preconditioning before they can really appreciate and accept these concepts; that you study these principles for at least a period of four weeks time. This would indicate: Number one, your sincerity; your desire for spiritual knowledge and secondly, it will provide you with what we call the frequency principle; that it would place you in tune and in touch with these Higher Minds who live in another dimension.

"Much work is done in this way and a certain degree of mental awareness is provided but the big work that is done is with the obsessions; the negative thought forms that are picked up and which are really a part of every person in this world. Many of you newer ones will be in for a great deal of introspection when you hear the nature of the feedback that we, as students, provide in terms of our realizations of what we have recognized in terms of our past, and how the past has been the basic block for our integration in the present.

"The basic philosophy (we all know this and I am only reiterating) is that we have to discover the cause of our malfunction for whatever mental disorientation or imbalance we have, in order to discover how to effect an equilibrium. Everything is the mind. If the mind is disoriented, if the television screen is out of sync, we have to go back to the operational function of the transistors or capacitors or whatever caused this malfunction. We, in a sense, must go back then to the true nature of ourselves which is our energy body, mind, our psychic anatomy.

"So with that mental process instituted, this should be more or less related to anyone who would like to come to one of these Unarius meetings and you have the responsibility to orient these people who must, themselves, have some preconditioning through studying the text which is preferably the "Voice of Venus" which is the introduction to the Unarius Science.

"So we will get into the basic format of what we are going to do tonight and that is the re-engagement of a past that has been with us for many hundreds of thousands of years. Now this is the first possible upset with those who are not aware that each one of us has lived thousands of lifetimes, spanning thousands of years and even into the hundreds of thousands. We didn't appear here in one lump sum; we took with us the sum and substance of what we had already accumulated from living in other time-and-space periods in England, France, Spain, in other parts of the world and on other planets.

"Here comes the subchannel; maybe he will take over. I am making a little introduction to familiarize you with the basic tenure of tonight's program. I am glad you showed up."

Vaughn: "We are going to have testimonials tonight, so I will give my testimonial first about what has been going on. I just read the article in the latest "Omni" Magazine that really shows very definitely a very big part I took during the capture, torture and subsequent demise of Dalos. The magazine contained an article called the "Bionic Brain". I won't read the article, it's too long, but what I was doing during that time was trying to have Dalos confess, not necessarily to secrets but to confess to the rightness of the Empire's (my) position. This is what I have been reliving many lifetimes since this Dalos period, and I have tried to control and manipulate these Higher Beings so that they would go along with what I thought was right.

"In the latter days of Dalos' torture, it was found that this Being could not be subverted; he could not be tortured into coming into the fold of the Empire. It was thought by we who led the Empire, when Dalos began to journey into the planets controlled by the Empire, that we would sway this highly intelligent Being and in this swaying we could encompass more planets. The Confederation of Light was of course under the auspices of the Higher Beings. We felt that if we could do that, then we would prove we were supreme! Well, as we have said, the Being would not sway and I could not believe nor did I at that time see that this Being was a Spiritual Being, a highly evolved, Spiritual Being. I looked upon this person simply as a human being like myself and also very intelligent, (not necessarily like myself).

"I felt that the Empire could do with a person like this, so it has been my realization in the past week, that I am the one who conceived the whole plan! I am one of the leaders who conceived of this plan and set it in motion so when Dalos arrived at the inner circle of the planets the Empire controlled, that this instrument could be implemented if persuasion could not take place. The Confederation at that time was coming in contact with the Empire, and Dalos was hopping from planet to planet; from the outer rim of the Empire's planets to the inner rim which held the seat of government. This took quite awhile to do and we knew this Being was coming to the central place, so we took these steps that you all know about, contriving a force field and implementing a computer-type torture device that incidentally was used on many people, not just on this one soul during the years of the wars.

"Progressing a little further, it was realized that this Being wouldn't crack under our manipulatory control, so we decided that one last effort would be made. If we could not persuade and dissuade this person from the true course of life, to go the way of our Empire, then we would actually take the physical being apart and reinstitute bionics so that manipulatory control of Dalos could be generated from a machine. The article in "Omni" magazine tuned me in explicitly to the way we did this.

"It talks about the computer being attached to the brain and the brain communicating with the computer by means of which the brain could be stepped up to very, very rapid paces. I tuned in to the fact that the computer could do this to the brain under certain conditions. And in stepping up the brain, the basic defenses that are instituted in the psychic body would break down and these impulses would override the natural information stemming from the psychic anatomy. It was our intention to make a person think what we wanted him to think. This article in "Omni" more or less, scientifically explains the viability of this concept; that it actually could take place, even today with our very limited science. If you want to read it, you might become attuned to that time and to our wicked efforts. It has been a big help to me.

"Sunday, I tuned in to this whole thing about bionics as I was standing in the back there watching the monitor. That monitor was, to me, that manipulation of the dials and turning the lights on and off was really, to me, the bionics attempted by us to get Dalos to look as if he were complying with what we were trying to transmit over the airways to the Confederation and the planets that were wavering and not involved in the Empire - those who might listen to Dalos. I reacted all day Sunday with this energy in mind and with this reaction, I know for sure that I have been fighting the Light from Day One! And this fiasco last Wednesday was my attempt to deride and destroy the Light by stopping it from its true course in this dimension, as I have always done from my own negative past.

"It is the truth, that if you don't understand something, you have got to destroy it (in your own mind), and if you can't destroy it, you are going to attempt to place it up on a pedestal. So this was my realization: There was an attempt during that war period of the Unholy Six to torture and then to institute bionics to transmit this hoax and this falsity to the many inhabitants throughout the Empire and to transmit this information to the Confederation. When this failed, we simply turned off the machine, and it did fail, miserably!

"Sunday, in my estimation, nothing went right. Well, it didn't go right the way I wanted it to. And what did I want? I wanted to turn Uriel into a robot like I am, because I don't understand.

"Last Wednesday, I denied what she was trying to point up to me because I didn't understand and I say today, 'I still don't understand!' That's why I am here in this dimension: because I don't understand. The Science is here and I am going to attempt to understand but in all these past lifetimes and even in this lifetime, I really thought I understood something, however, during this cycle, I have come to realize that I don't understand and therefore I am in opposition to that understanding with the Infinite, until I can understand; that is, I must study and overcome these hard-nosed, egotistical pasts of my own. Until I achieve this understanding, I will be in opposition and will still revolve in this Earth world as a miserable wretch.

"When all our efforts failed, we destroyed her. We destroyed Dalos just as we destroyed everything we got our hands on, just as we destroy everything we get our hands on today. We can't do anything different until we learn to understand the Infinite and become likewise - as these Higher Beings. I don't know how long it is going to take me, personally, but I am going to make the effort. So the rest of the evening we are going to leave open to anything you all have to say.

"Incidentally, I want to make it clear that I see now that last Wednesday, my reliving was to try to incite the Confederation and the Empire and anyone who would listen, against Dalos. I take back everything I said last Wednesday. I hope every one of you understands it as a reliving. If I were in my right mind, I would have said none of those things but that was the reliving on my part: to attempt to reinstitute or reinstigate my past. I would also state that it was a failure only in my eyes, as the Light never fails and that Light has never failed us. We however, have failed it time and time again. Even during this negative cycle Sunday and subsequently, I felt very despondent and depressed, yet the Light won out as it always wins out."

Jeff: "The lifetime I lived during this period when we captured Dalos (or more factually that we trapped our own psychics, which is more the truth), I would have been better off being aborted. I have some realizations about being a very busy individual; I was very busy destroying myself. I worked behind the scenes, planting the fake Dalos on Idonus and I was involved in many other projects, but basically, I was a behind-the-scenes man.

"One of the things that seemed to pop out of nowhere while we were filming, was the idea that we had conducted some kind of experiment to prove that the Orion Empire was sophisticated enough in technology to have people incarnate at the time and place of their choosing. Once we did that, documented it and proved it to people, the job I took on was to extend that further and say, 'If you die in battle in serving the Empire, you are guaranteed a lifetime on the pleasure planet of your choosing.' It was like winning the sweepstakes. But of course when a person commits suicide in any way, shape or form, he hasn't any more integration; he has lost all alignment. Then, without the help of Higher Beings, he is in very bad trouble, indeed! Of course, people are not in any mood to be helped by the Higher Beings if they have just committed suicide with the idea of spending the rest of their next lifetime totally indulging themselves. So there was a tremendous amount of damage done there.

"This was brought inphase when Uriel kept looking at me and when she said, 'Kamikazie, kamikazie,' when I was wearing those batteries around my waist. There were many people who died at the time, thinking they were going to get to a sensual heaven, who never made it. This misconception has come down through the ages in the form of the warrior's death and was very popular with the Nordic tribes in Germany and so on. On the same day of the filming, I was talking to the subchannel about the beliefs of a Roman soldier. They thought the worst thing that could happen to you was to be sent to Germany because those crazy sons-of-guns would ram themselves onto a spear so the guy next to them could get by and get at whoever was holding them. There was just no way that we could figure that out. This is a very vivid memory with me. The reason it is so vivid is because it is part of the propaganda I handed down. I feel that it must have regenerated down to a time in the Middle East where there was a cult from which we get the word 'assassin'. Assassin comes from hashshasin which means hashish eaters.

"These people would be put through some kind of crazy religious ritual over a period of years. Every once in awhile they would be given a certain amount of hashish to make them stoned out of their brains, then we would bring in several buxom young ladies and they would figure they were in heaven. When they would come out of it, we would say, 'Okay, if you die on your mission of assassinating someone, you get to go to that heaven. Isn't that neat?' So they would willingly die - they were totally crazy! They wanted to get killed doing that sort of thing. I saw that in a movie about Omar Khayyam and really reacted to it and didn't bother to look at it. I don't know whether I was in that cult or not but I know that the energy from that cult originated in part with the work we did during the Orion Empire.

"Sunday night when I was reading the proclamation of welcome for Uriel, in point of fact, what I was doing was reading charges against her! I think I realized that as soon as I opened my mouth. Earlier in the day I was talking with one of the students in the audio department on how to strip the audio off a video tape and put it back down. The problem is that you don't get lip sync because the speed of the tape recorder and the speed of the video machine are a little off. So we wanted to build a machine to make it in sync. What we were doing was trying to figure out how to manipulate Dalos in that lifetime; to mechanically move his mouth and get a voice track that we had edited over a period of time. Then whenever he spoke, we would take a word from here and a word from there and put them all together in the form of a confession or something that was detrimental to Dalos and the Light.

"So when I saw Uriel Sunday, moving her mouth to the sound of words (in our voice-over tape) and that awful blue light with the hood over her eyes, I knew right away what was going on. I was able to get past it enough to appreciate the energy, the future that she projected through to us. It was absolutely, tremendously beautiful. When I saw all the wires that were hooked into the dress, I immediately tuned in to hooking electrodes all the way up her spine into the base of the brain so that using various levers and dials, we could manipulate her to move her lips and hands and various motions of that sort. She got into it and moved like a robot and I said, 'She's really tuning me in to it and she's not going to spare me with this one!' So that was pretty horrible.

"I also believe that in the lifetime of my reading charges against her, that this has come down in many ways in various lifetimes when I have accused her of various things. The military frequency that I incepted in that lifetime or made negative, I have carried with me ever since. I have come to realize that I am only slightly out of uniform at any given time; many of the clothes I wear are actually uniforms. I don't dress casually because you never know when you are going to be on parade. I really believe that after all these years of trying to get Dalos to submit to our will, the way I, personally, rationalized it was that this person is a freak of nature. There is something unique in this person's genes, chromosomes, hormones, brain structures. There is something that nature has done that is very unusual here and if we take him apart, we can figure out what it is. Now, I think that seeing this thing of the brain, after we had disposed of Dalos and he may have been dead when we did this thing of trying to get a lip sync, that we took him apart, cell by cell, to see what it was that made this person tick. If we could figure this out, we felt we could build a whole army of them. So that is rather gross.

"I would like to express my appreciation to Uriel and the Brothers for the phenomenal patience that has been expressed lifetime after lifetime and that I see them in a new light now that I don't have to block out that area of my consciousness. The phenomenal mission that they set out on is just words and when somebody says to me, 'Nineteen million years ago these Beings set in motion this vast plan', (which is the truth), and when I see it in terms of my own evolution where some hundreds of thousands or millions of years ago I did all of these terrible things to a beautiful Higher Being with all the malice that I could conjure up and that Being, from then until now and in the future, has nothing but love and wisdom to offer in return, that is the true meaning, I believe, of turning the other cheek. I am going to strive to become one of those Higher Beings."

Mike: "Sunday, I was totally inphase with this specific past and I am totally agreeing with what these persons say - that we were trying to make a robot out of Uriel. My specific part in that life was in the making of bionics for her anatomy. In this life, I was involved in making and designing the stand, the hoops and tying the chair down. But what tuned me in was that Sunday, Uriel said she felt like she had no spine; she had no support to her back. A few days before that, she wanted me to make her some kind of support on the chair so she would be able to lean against it.

"Well, for some few conscious-mind reasons, I didn't. This shows that I was working against her. Right at the end, we were having problems keeping the huge collar up, so what did I do? I added a copper tube that ran up the back of her spine and tied it there so she would have a little support in the back of her collar. In reality, what I was doing was adding the final electrodes in her spine so she would be robotized.

"All during the Cosmic Flame presentation, I saw nothing except her agony. To me it was a total disaster - which it was in that lifetime; a total disaster. I also had all the guilt and the fear in this present lifetime from her going through so much torture. So that's proof enough to me as to what I was doing. I hope I have enough strength within myself to carry on and progress in the future."

Mal Var: "The huge control board I have been building upstairs is obviously the controls to operate Uriel (Dalos), to manipulate him and to cause reactions in Him so they could be filmed. I had a workout on Sunday when my brother came down to see me and I was very happy he came. I showed him the control board and the first thing he touched was the control board. He goes like this, (Mal Var goes through the little twisting motions from one to the other), touching the plastic knobs and all the buttons fall off as he touched them, with a 'clink, clink, clink', down to the floor. I didn't get particularly upset about it; I just cleaned them up and said I would glue them on later. I still felt a little negative and said to myself, 'All I have to do is glue them, why do I feel this way? It is so easy to fix.' But I still felt this block come on and I felt very uncomfortable. Maybe if I would take an hour, I could glue them together. I wanted to get the control board back together again really quickly but I was with my brother and didn't bother with it but I felt this negation within me.

"That night, during the presentation when Uriel was pulled out on the platform, she wasn't smiling in my consciousness and I felt very, very disturbed about it. I said, 'The whole thing is ruined! She doesn't look happy so how can I feel uplifted if she seems so miserable?' I kept wanting to scream out to her, 'Smile, smile! Uriel, smile!' I was very negated by it and I realized that my control board was broken and there was no way I could manipulate her; I had no control over what kind of expression she was doing! I was filming the video and I said, 'The video is ruined and this isn't going to be much good. So I just sat there and filmed it without much consideration because who cares to look at her if she doesn't look pleased and proud of the great accomplishment?'

"So this really was brought into play for me, that I definitely operated this control board over her, that I had constructed. I don't feel that I constructed the electronics to it but I do feel I constructed the layout - the convenience of it - how convenient it was to operate these knobs with the convenience of sitting. I made it convenient so it could be operated correctly. It broke down and I was the operator and felt I had no control. There was nothing I could do (in the present filming) because you don't scream out to Uriel, 'Smile!'

"So I felt all closed in; there was nothing I could do. She is the director, I can't direct her. The thing was ruined and I really fell down, but realizing what it is, now I feel a lot better."

Hugh: "This has to do with what Tom said: that we thought Dalos was merely an intelligent being and we didn't understand the true development of this Being. We had developed the technology to where we had space travel and all kinds of electronics, computers and so forth but we didn't know the actual soulic development of this Being.

"A couple of weeks ago it was brought out that there were many of these things we didn't actually do to Dalos in a physical way but that we projected these illusions with the computer for Dalos to experience. It was brought out, that in anyone's evolution, he has abilities and of course Dalos had worked them out but we unblocked the cycle to where Dalos could see these things again from the past of lions and snakes or whatever was projected to him. But what we didn't realize, in our ignorance, was that in Dalos' soulic evolution, he had been tested for illusions and pasts at that point.

"Some people might know that a person, in a certain point in his evolution, is tested for illusions and he learns to tell the difference between the illusion of this world - what is and what isn't. They are not fooled by a chasm that is not there, or a lion that they know is not there - that it is but a hologram. Dalos knew these were illusions but we didn't know he knew this. We just kept on and on and couldn't figure out why he didn't break. I am saying all this because I was involved in all that; in the gross ignorance of it and helping with it.

"I think I have already stated that I helped to do these replicas of Dalos with plastic surgery; I was a plastic surgeon then.

"There was one other thing I wanted to bring out. At that time I had the most elaborate equipment, the finest equipment; everything was at my beck and call - tools and the finest of everything. I was always bothered in this present lifetime if someone moved something and it was not on hand when I needed it, or it was bent, or if someone were to misplace my tools I would really react. I know the reason for that reaction is that I don't have the tools now to do everything I want to do but I was given all the tools I wanted back then in that past and used them negatively. So I had to learn this lesson."

Michael: "A few weeks ago we watched a movie on cinemorphology and they took brains and froze them and sliced them into micron thick sheets and photographed them. By doing this they were able to see the exact cellular structure of the brain. After I went home, I thought that in Lemuria we took full bodies apart and did that. That is how we built androids. I just realized that eventually, as Jeff has said, since we couldn't figure out what made Dalos so different, I think that is what we did. I think we found the particular things that controlled certain responses and wired up Dalos to an instrument that could send out those minute electrical charges. I know I was involved in this and I was especially involved in the public relations aspect of it, making sure that nobody suspected anything because when the Cosmic Flame was shown, I was sitting where I could see the left side-curtain open. It was left open after Uriel was wheeled out here and you could see Steve Anderson at the panel. I said, 'Dan, close that curtain; it is going to ruin it if they can see what is happening.' That is what I was reliving having done - polishing it up as if Dalos was actually sitting up there confessing and joining the Orion Empire."

Frank: "Sunday brought out further proof of my feelings with this involvement with the tortures of Dalos. My concern for the success of the event was so strong because of my guilt for wanting to be successful in controlling Dalos and getting out the propaganda to strengthen the Empire's position and dilute the principles and teachings of the one called Dalos, at that time, and that every little detail seemed to be my concern. As you know, I am one who kind of touches base; my involvements have not really been strong with the mechanics. I just kind of touch base sufficiently enough to get to know who is who. This basically represents what I do: contacting leaders; people who are in charge of affairs as well as those who are supporting them. I keep the communication lines going. Some of these people are a little troubled by my moving about like that and rightfully so because the liaison position I had at that past time was obviously being in contact with some of the leaders who were in competition, including myself, for the position of Emperor.

"When dealing close to the Emperor and with the others, we all look out for 'who is going to do what to me next?' We have all done some pretty serious things to each other which has come out in previous testimonials which I have fully realized. We would take somebody's life and think nothing of it. We thought only of our own particular desire to move forward and be strong in the Empire.

"The great concern for Uriel, which again proves my involvement, was interesting because she had concern for me as I was assisting with pushing out the platform; she was concerned that I might get hurt. Her concern for me in the present, was a reflection of my lack of concern in that past lifetime for Dalos in the positive side at present. The snap, crackle and pop you heard and the odor from the electronics, the light that was brought up earlier (the ugly blue light) and the replay of the audio with the syncing of her lips was to perfectly attune us to what it was like. We were trying to perfect that illusion to where we could get across to the viewer the erroneous information and paint Dalos as being an imbecile, not coherent and with nothing substantial to go on. That was one aspect and of course we were trying to perfect it to the point where we could really make it look good and have Dalos saying things in favor of the Empire. This was a very difficult thing to do and obviously we did not succeed and this is what caused all this looking into Dalos and the horrible end of Dalos.

"To be a part of that, or endorsing it, or having any part in that at all, is a very devastating thing to one, therefore my guilt is so heavy that in trying to do the right thing for Uriel, I am a little pushy at times and I am not as patient, even though I try to be. But inside, that energy goes out to try to right things and I have a tendency to push people too hard. I must remind myself to take care of this guy (pointing to himself) and not push things off on anyone else that is not their business. Inside, I have premeditated many things and I am trying to do the right thing for the Light and in the process of that, we can get a little off; one can get a little pushy. Therefore, I apologize to anyone who may have received energies from me; it is just a little overconcern on my part. I need to study a bit more and realize my part more and take care of that and let the other people take care of theirs. So that came to a head tonight too, which is a good experience for me.

"But most of all, I am very pleased that we got through last night. My concern was heightened when the heat was hitting and obviously so with the crackling and popping, Uriel was very uncomfortable. I was standing by, ready to do whatever was necessary if I saw any spark or flame that looked like it was going to go afire. I was ready to jump right in because, my gosh! It was terrible what we did to Dalos in that time."

Christine: "This is a big one for me. I just finally realized what I was doing up here, clowning around last Sunday night. For the people who didn't know what was going on, I was impersonating Uriel, only I was trying to make a fool of her. She put me in this position to do a little acting skit. I was supposed to come from another planet and say, 'Well, I snuck in here and I hope nobody minds but I am getting information to find out what is going on here.' And Uriel is so wise and so intelligent, she can see my past and knows that which I have to work out. She got me this dress and it 'happened' to be just like hers, and she let me know that. She said, 'I got you a dress, and I liked it so well that I had one made just like it.' Right there, I knew. So I have always wondered whether I had impersonated Uriel and here it is, coming out! So I waited and waited until the last moment to hem it up. I didn't want to wear it but I knew I had to go through with it. So I wore this particular dress which was a duplicate, in another color, of what she wore Sunday night after the filming program.

"She asked me to do this Chinese skit, which is different than coming out here and acting like myself; it is acting like another person. This put me in the position of impersonating another person. So there I was, getting laughs from the audience and acting; I am sure I clowned around and acted like Dalos in that particular lifetime; I am positive of it. From that time, it has shaped my personality to be an introverted person and not letting people know what is really underneath. Ever since I came into Unarius, I have been trying to get rid of this introversion, but I know what this is now. It is this memory of Dalos in my psychic that I am trying to hide.

"Also, what I have been reliving is putting the electronics into Dalos' back. Another student and I have been working on these sequins, not on this particular banner but on another one that is in the back. These sequins represent the electronics that we put into her back. Ever since I have been working on that, I have had a horrible backache. I didn't know until yesterday what it actually was. In fact, I was not even going to come to the Sunday night performance, my back was hurting so much. So, right there, it is plain to see that I am really guilty of having hooked up the electrodes.

"Another thing: During the film that Mal Var made, we were all clapping for the planets that would light up and shoot by, one by one - El and Donatus' planet, and all the different planets. At the third planet, I realized we were actually clapping for the planets we had overtaken and conquered, cheering for our great destruction."

Roberto: "Ever since I came to Unarius, I have had this great obsession and really great negation. It has been my fear of Uriel. I have overcome it to a certain extent but lately it has regenerated due to the fact that this one cycle is one of the mainstays of this fear. As proof of these negations in this Dalos cycle, I have been involved in viewing the movie that Vaughn brought to us about the brain. As soon as I saw the picture on the screen of the brain, all analyzed and cut into certain sections I said, 'This is Dalos' brain.' I didn't have enough information and just put it aside and did nothing with it. I noticed that after viewing this movie I became very involved with building the Cosmic Flame.

"I began soldering this sequence of wires that went to the lights on the garment. I stayed on it one day and night saying, 'This great obsession has to be done with, and the way you guys are doing it, you are not doing it right; you are soldering wrong. You are leaving cold solder behind; only I know how to do it! I'll do it and I don't want anyone else to do it because you guys are going to do a sloppy job and we are going to burn her up. This is the way you have to do it!' These were my attitudes and I kind of slowly took over the soldering or assembly section.

"In addition, Uriel gave me an opportunity to play a little rendition during the presentation last Sunday. As I played what I believed was good, snappy music for the Flame Presentation, I received a statement from her. She said, 'Do you know how to play any music?' Of course that just about blew me apart. Ever since this cycle began, Uriel has been trying to get me to play marches. That is the last thing I want to do - play a march. I am a classical writer or musician. Do I know any marches? I just wipe them out of my mind. I can't conceive that one-two beat rhythm; it just doesn't register with me. I don't even like to watch people march because it is robot music. That's what dawned on me; it is robotizing music!

"So I realized that I was one of the assemblers of the bionics that went into Dalos and how could I say it? I had a negative pride that my work was going to be perfect and anybody who was above or below me had to follow me, because if I told them how to do it and it worked right, then I would get the credit. It turns out that I had a great deflation last Sunday because I expected to be called up to be praised and Uriel, with her infinite knowledge, did just the opposite. She praised everyone else but me and I got the deflation because I was working from the ego to begin with. So the ego deflation came about and . . . I forgot to mention that during the presentation I was running back and forth between (audio electronics), VTR, out here and the back room. I was checking to see if the VTR was on and the fuses were off - I mean, 'went off'. The VTR's fuses were off and the things were going on. I was responsible for the audio for the first part of the program, manipulating the dials so the audio would click. So my realizations on this are not only that I was an assembler but that I worked on splicing the words of Dalos and syncing them with the device that I helped assemble to do the work of these phrases that I would put together, to which I would have Dalos move his mouth.

"This is why my music was not integrated; this is why I react to rule-types of music, because this is what I fed in and used to make the perfect robot of Dalos! This is what I have: a big realization for me to cope with!"

Channel: "Speaking of giving credit where credit is due, is Thelma here tonight? We want everybody to know that Thelma has contributed $700.00 to Unarius. That is how we work out karma - when we can see fit to give of our income and contribute to the others' overcoming too, because that is how we overcome. When we can get up here and speak the truth about ourselves, this is one of the greatest opportunities we could have. And we should really continually thank the Brothers for this opportunity. There are a few more commendations that Uriel would like to have made. One is about Mr. and Mrs. Hook; are you here tonight, Decie and Gordon?

"Uriel wants you all to know that these dear ones contributed this fine organ to Unarius; not only did they contribute it for the use of the students and the various productions, but they also have recently purchased a wonderful table saw. Mike Wilson helped with the cost, but the Hooks have been really dedicated lately and we would all like to commend them for their efforts! This shows on you two, because you are really changing like night and day, just during the past few months! I would really like to commend you both. We all appreciate your efforts, and to all - you students who are making the effort to contribute by reversing your negative karma!

"Mike Wilson not only contributed his money, but he also took a very active part in gathering the material and helped in construction of the Generator so that it could be moved forward in a positive way, regardless of the existing negative energies. I would like to give Mike a hand for that, and also for the fine saw. You know we are making a lot of things lately and we need this type of equipment since we are going into the theatrical or filming business and we are needing various equipment.

"Gordon Hook is such a great artisan in this sense; we really appreciate his abilities and that he is sticking to his efforts to work out his karma. Mike Wilson has the same abilities although physically he keeps claiming that he doesn't know anything. But this is the way it is when we work from true inner guidance, and what we have set up and conditioned our psychic selves for, in the past, to work positively for the Light.

"While I am up here, I would like to also bring a few more points out of my own past reliving with this Dalos cycle that I conveniently forgot to mention. To prove my past, I conceived of the way to put this Cosmic Flame Configuration together and I was continually trying to explain how to put this together to the various participants who were building it. So I believe that by this continual attempt to explain to others how it should be done, I conceptualized it and let the various engineers, or badgered the engineers into building this device! Also when Stephen put this stage together, before the curtains were installed, I immediately saw it and walked into the middle of it. This was about three weeks ago when I walked into the middle of it, when it was just the wood - just four corners of wood framing and I suddenly felt very heavy.

"This shows that everything you tune in to is psychic. There was nothing there, no constricting device to really tune me in. There was just the square shape of it, four 1x4 pieces and I tuned in to the force field room that was built. I became quite upset at Stephen at the time, saying he was building something unnecessary. (Stephen was setting up a stage set for our filming.) But more importantly, after I walked into the force field room that was supposed to lock Dalos in, I locked myself into that force field room. Ever since then, I have been in that force field myself because of the concept that what you cast upon the waters, returns to you tenfold. I was trying to lock that Being in, and here I was, getting locked in as the backflow of the negative energies. I felt really unhinged and locked into the negative past here for a couple of weeks and I couldn't put my finger on it, even though it was very horrendous.

"Another part of this reliving is that when I first came to Earth, 800,000 years ago from Planet Pluto, it was the preceding lifetime before the Unholy Six life, that I had incarnated into a technologically advanced civilization. They were exploring the solar system and I had a receiving device implanted into my head so I could communicate with the people on Pluto who had a base there established to communicate with the inner planets. This implant that I had, displayed itself physically in this present lifetime, as some of you know. I had the realization today that this is what I was trying to place into me - the electronics or the bionics that I had placed into Dalos' head! This is why I went haywire when I came down to the Earth plane. The device wasn't negative or to send raybeams out or anything. All it was in the physical was a communication device that picked up my every thought and visualized it for the comrades who were monitoring me on Pluto and there were various others on other planets in this system.

"Well, the whole thing went haywire. I reverted and became very negative and tried to destroy the tower in that time. Since then I have been locked into this Earth plane and haven't been able to incarnate on any other world. But this is proof to me. What more proof can a person have than to visualize the various things that happened to him in his daily life that not only represent but are so blatantly explicit of these pasts? How can one deny them? And I was denying them because this past is so horrible and this past is so gruesome, grotesque and ugly and black that we don't want to see it. I know I don't want to see it, because when we really come to grips with instilling this energy into our own psychics, it is a horror movie replayed twenty-four hours a day in our psychic until we can get rid of it! It is the demons and the devils we have linked to and hooked ourselves up with, so much so, that those devils and demons have become us and fight and destroy the very things that can help us.

"So, that is what happened to me. I locked myself in and that is why I could not understand what was going on when Uriel was trying to help me understand what is negative and what is positive. What are you doing that is not in connection or harmonious with the Light? Well, in these times when we are locked in, reactionary-wise, we don't understand; we cannot cope with the positive information; it just comes in distorted, meaning that when it gets into our psychic, it is distorted and then it comes in. So even the Light, the Power, the Truth and all of those things get warped and distorted by our own negative pasts and look illogical and unreasonable. Only when we can unwarp and untwist and get rid of this relentless, negative force pressuring us to do these atrocities against ourselves and our fellowman can we see the reason and the logic behind the progressive way."

Jeannie: "I finally feel the reality of this lifetime. I can definitely relate to what Tom (the subchannel) says about being locked in to the negative past for quite awhile now. I was just given one clue after another of my participation in this, but I have chosen to not really look at it because I was so afraid to look back into this awful negative past in which I am so trapped.

"A few weeks ago I wrote a testimonial and sent it to Uriel about some realizations I had about implanting bionics in people. I had gotten these clues from the patients I work with at the doctor's office, for every time I hooked up the EKG machine to these people, they constantly gave me one clue after another. They were telling me, 'Oh, it looks like you are separating me into different parts. It looks like you are going to electrocute me. What are you doing with all the wires? It looks like you have a control device there,' one comment after another, after another! Then soon, before I took a patient into this room to do an EKG, I knew they were going to say something to tune me in to this past, but even looking into it in that way, I still was blocking it off really strongly.

"I reacted very violently to the film Vaughn showed on the brain, and in this present lifetime I have been very interested in the fact that we supposedly only use a certain percentage of our brain. I read a book called "The Brain Revolution" which I found very interesting, telling of the different functions of the brain. I believe that in that lifetime I tried to stimulate regions of the brain that were not used. I believe I was involved in trying to find out what was so special about Dalos and why, after all this torturing, Dalos still wasn't affected. I really thank the Brothers for the help they are giving me. I am glad I am finally feeling some of this negativity so I can change it around. Thank you."

Dan: "Last Wednesday when the subchannel was having his big workout, he asked me to moderate the meeting. We had a talk and I ended up talking him into coming up here and attempting to moderate, but that was really a reliving of a past (or maybe many pasts) when I wanted to usurp Uriel's position and be the Number One, positive, powerful person, leader or god, and in this particular reliving, I saw Tom as opposing her and this was my chance to exert my own influence and gain position. Even in the back of my mind I was thinking of betraying him and taking over from him at some later date.

"I have attempted to do this in the past but I have never succeeded in it because I didn't have the power or the understanding. Of course, if I had, I would never have tried it to start with, but that is the great extent of my ego.

"During this time that the group of us executed Uriel and I knew she was an important person, it was a sort of behind-the-scenes thing. I was the ax man and I knew that this person shouldn't be axed but I didn't care because it was my job and I wasn't going to let something like that keep me from doing my job and lose my position, no matter how small it was. I feel that there was some sort of investigation held later as to how or why this happened. This person should not have been executed, it was said. The buck was passed down and it seems to me they were trying to blame me for axing Uriel (Dalos). This is what my reaction was, 'Well, I am not going to let them get away with this! They are not throwing the blame on me; I was only doing my job and it was a joint expression or display.' That was the reaction to that.

"Then when she was up there in the Flame display, the tableau we all made, I relived my part in the Dalos Cycle also. She looked like a robot to me and I helped a little with the electronics. But I was the one who got up there and announced it, I told everyone to, 'Hold on for a few minutes; you are going to have the thrill of your life and you will see Dalos in person,' which was the robot thing, so I really was up to my ears in it!"

Crystal: "I am guilty. I have been reliving all of this. I am not going to say I understand it, but I have only just begun to accept the things I did in that time. I am just frozen with fear and guilt and fear of being exposed. I know I was involved from the very beginning in the whole plotting and planning. I knew what was going on as I was involved as a worker on it. I really got actively involved when the robot was made, and all of a sudden, one day when I came in, I began to work very closely with David Reynolds, like a very close assistant. The main thing I can be aware of now as to my feelings at the time, are of not being able to crack or break this person with the anxiety, the frustration and the determined efforts put forth to try to break that individual. When we could not do so, then the fear of not being able to overpower him left me in complete torment and frustration within myself and with a desire to institute more torture upon this Infinite Being, Dalos.

"Besides knowing everything that was going on, I was involved in going on the outside and pretending that it wasn't going on. So, knowing of all the horrible things and all of the lives that I have had to pull this off, has created a great pressure of being exposed; it is a great fear within me and it all is heavy. It is I and what I have done at that time; every thought, every action and deed I create in my own psychic I am reliving.

"I had a dream about two years ago and it was of this huge electronic monster. I saw a picture of this woman who I knew was I, bound and determined to break the back (those are the words) of this creation that had been created in the past. At one point in this dream there was this huge spine; it was like a dinosaur spine and this person was telling me step by step how he had constructed this spine and it all had to be corrected in the future. There was another point in it where I took this spine and cracked it and all these electronics came out. I put the analogy of this experience in the time of Lemuria when everything was on electronic control, but this spine torture goes back even further. I was the one who has been working on these little spines here (in the Cosmic Flame) with Chris.

"I know I am now correcting that. I have been

very involved with the construction of the robot but the main thing in my psychic is the fact that I went out with the people knowing everything that was going on and told lies exactly opposite to what was really happening. This was public relations for the Empire, I guess but I was there in both places. I am confused but I am asking; I want to be open; I want to face this past and become free from what I created within myself. I can't get up here and pretend I understand everything I did, but I know the Brothers will give me that which I need to know when I open up and want to really learn and accept what I have done."

Mal Var: "I just must tell this; I have been boiling over. When I sat down from having given my other testimonial, I was still having realizations and projections from the Brothers. I still have a great deal of karma with Uriel from this lifetime; there are a lot of various things for me to look at. This control board was also operating the electronics that were in Uriel; I do believe that would move her around. When the control board went down, I could no longer control her and the mouth-sync thing developed trouble. When I was looking at it the other day, I could not stand looking at it; I was seeing my past. That's why I wanted her to smile when she was presenting the tableau because I couldn't stand looking at her. But the further realization presented here was that I did not press this particular control, somebody else did, perhaps my brother and it malfunctioned and was wrecked, but I built the control board.

"The reason I say I didn't build the electronics in the control board is because I feel some very heavy karma with the electronics and the control board and I feel it malfunctioned and I see Dalos (Uriel) moving and jerking about. Then I see her mouth being pulled down on the sides from malfunction of the board and this explains the horrible electronic-type smell that was present and which I react to violently in this present lifetime. I have always had psychic reaction to it; it smells like burning flesh to me and it hurts; it sends physical pains through me! It is a physical charge through my body; it is a cold chill and it does hurt.

"In school, I used to arc weld but I had to give it up because of the smells that were emitted in arc welding. I couldn't tolerate the smell and I told the teacher, 'I can't arc weld because the smell causes a pain in my spinal cord.' No one could understand that because it was (is) a psychic thing.

"I feel that this control board was responsible for his eventual death - that I helped to kill him. The reason I feel this is because my car has gone through three generators and we all relate to this Flame as the Cosmic Generator. When I was working on my car with another student, we wired the generator up wrong and in a flash, the generator 'fried' and was gone, so I knew there was some karmic thing going on here.

"I also relate to a lifetime I lived in Lemuria which came after this. Uriel was filmed and she was eventually killed, then we made a robot of her and I filmed that also but then I was killed because I wasn't going along with the situation. Also I wasn't getting enough credit for the position. It makes sense in my consciousness that this control board had malfunctioned due to someone else pressing the buttons but I built it and it caused Dalos to fry with these electronics; afterwards a robot had to be constructed. This is where the robot came in - after his physical body was killed, so the robot had to be continued as long as we had developed this bionic system. I don't know if anybody relates to that but I feel this to be true."

Keymas: "I relate to that too because I felt she was being burned to death. As for what Dan said, I also felt very strongly that she was being executed. I am glad that somebody said something about it to help me clear that up for myself. My reaction has been, 'Well, my control box works better, Mal Var, compared to yours,' which is the great ego I was involved in. I helped build the control box that housed the switches and electronics to cause the various functions on the flame construction. I did this with tremendous pride, more so than I have ever felt as I took a great, egotistical pride in this, although I tried to keep down the ego; I was blocking it off, I wasn't keeping it out. I did the soldering of the wires which I had to redo three times. Benno kept giving me a 'PQR' reject and I feel that I wasn't qualified in the electronic phase of this and was sent to another section because all of a sudden I found myself (in the present) in the woodworking section in the lab. So here I was making the box and doing it with more precision than I have ever worked anything with, redoing it and making everything fit just right. If it didn't fit just right, it had to be redone.

"I feel that in this past lifetime I didn't care what happened to Uriel; that wasn't my concern. All I cared about was my position and what I could get out of it because I wanted to manipulate the dials on this box! I felt it was my project as I put it together. I put the switches on so I should have been permitted to turn them on and which I didn't get to do. Well, I was a little resentful that I didn't get to do this. I didn't care whether or not this Being was put to all this torment and to eventual death; all I wanted was to have the position that offered more prestige than to be working in the factory section or sitting behind the control panel, manipulating dials, I wanted to push control buttons!

"I saw something in my mind the other day that somebody might be able to help me with, if they have experienced something similar which was that I visualized Uriel dying early. She somehow had died or was killed in an accident and her cycle here was finished but the mission was so important and so crucial that she remained in an isotopic, atomic form where consciousness was somehow monitored through her. I still don't really know what lifetime this was from, but I passed it off and remembered it later on in the day that this was somewhere in my past as a reality and I saw various students going up to her and saying, 'Uriel, is that you?' And some of us were fearful and she was saying, 'Don't worry, I am still here; I am just in a different body. We have got to carry on with the mission.'

"It was as if the Brothers had projected a force field around her house to keep her in a more fourth dimensional state so she could exist this way. I am still very blocked off with it and have been waking up with headaches every morning and I have one right now. I have been trying not to take aspirin because it's just blocking out this negation. I know sometimes it's necessary but I have been trying to hide it; it is too easy to take aspirin and have no more headache but it comes back like a boomerang, and I can't get rid of it until I objectify the cause of it. There is something horrible within me that I have to get out."

Stanley: "As far as I know, in this lifetime I have been in this cycle off and on for at least three months beginning when I built the vortex for the Venus skit when we enacted the Seven Centers on another planet. As far as I know, in this lifetime I was actually born to work on this project - to be a technician. I remember when I started reading "Brave New World" in the beginning, the book talks about how embryos were conditioned before they were totally physical. They were conditioned to be cotton pickers or whatever and I believe this is the reason I read this part - to tune me in to that time when I was actually born to be a technician on this project. When I first started, it had just been set up and was created and I began to work on the bionic parts and the plans of what we would do in the future were formulated for perhaps a hundred-year period.

"As I grew older and more people started to come in, I was resentful of them. It's like they were the 'young whippersnappers' and not aware of all we were doing. 'How dare they think they know more than I do, because I have been with the project ever since the beginning so they can't tell me they think they know what they are doing.' This is one of my reactions to a couple of the students and Roberto, for one. It's like, 'You are coming in here and think you are going to take over my position.' It was like a new technician came in and felt he knew more than I did. He actually did know more than I because we had evolved in our technology, so I didn't want to accept it but wanted to be the top honcho because I felt I deserved the position. The reason I know this is true is because of what Uriel said the other night when she (supposedly) praised us when she said that I sort of took this project under my wing. I lived this project as if it were my entire lifetime, and it was - in that past! Ever since this cycle has been inphase I have had this overwhelming desire to learn electronics. Almost to the day that this cycle came in for me, I started to build that vortex."

<u>Stephen:</u> "I finally got this execution thing a little clearer. It ties right in with the past

with Dalos; it is just a regeneration but it is closer I think, as it is stronger in my psychic. What I just realized will clear up something with you, Dan.

"Back in the earlier days of evolution people thought having executions was fun; they made a big festival of it. Whenever there was a criminal brought into the town, the mayor or whoever was head honcho at the time would have a day of celebration. I was involved in some capacity under the person who made sure these festivals came off properly. This particular one was an execution, so I had to think up some interesting device such as a stage to make the celebration a little more memorable or interesting.

"What I realized is that I had designed this device which was similar to a guillotine only it wasn't a guillotine because I feel that this lifetime was lived somewhere in the near East. It was like a Damascus Sword and it was used in such a way that when the executioner pulled certain strings, the sword would drop down. I had tested and retested and instructed the executioners in the proper way to do it and the proper dramatic style for the audience in the grandstand.

"Everything was in readiness and the prisoner was being brought in, who was Uriel. Now I had no concern for the person or being, whoever it was. My job was simply to set up the stage. I did see in my psychic this soft young woman who was being called on all of these charges, maybe witchcraft and such, just being an enemy of the public. I didn't care about that; there was no interest there. I cared if my set was going to look good and if the staging were going to come off well.

"So everything was in readiness and here are the grandstands with the leaders and the mayor of the city and I am sitting there thinking it really is going to look good. Then the executioner appears and the drum roll sounds, and the execution happens but it doesn't work! The executioner went off to execute Uriel but there was my anger at the executioner for not doing it properly, no matter if it was my malfunction or not which it probably was. But I feel this is why you resented me so much - because you failed to do your job well."

Dan: "My reaction during this thing was to Stephen because when we first did this curtain pulling, the thing fell apart and I said, 'Why are we doing it this way?' So that was my big reaction, and then when it failed, who got the blame? I did."

(Uriel: "I believe Stephen's subconscious prevented it from happening! The truth within him did not want her killed!")

Stephen: "Our opposition was purely personal and it did fail on Sunday night. I don't know if Dan choked up on it or not but he started dropping the curtain, then he pulled it back."

Dan: "We rehearsed this before we started and I said, 'Would you like me to tie this off here?' and he said, 'Yes!' I was so strongly involved in the reliving that I was even getting flashes of what I should do - raise and lower it. Stephen said, 'Tie it off.' And I tied it off and sat and said, 'I wonder why I ever did that!"

Stephen: "I am not talking about that curtain; I am talking about this first outside curtain."

Dan: "Well, I am talking about this first curtain too. To me this one represents the ax! So Jeff and Tom were reading the scrolls and I couldn't move the curtain while they were on stage reading. So as soon as they were finished reading, I went over and lowered it. What this represents to me is the reading of the charges, then the executioner comes up and then, 'Down with the ax!' "

Stephen: "I can't say much more than that! But also the harmonic is regenerated from Dalos because when the Flame was ready to start to go through its electronics cycle of lights and I came out in the audience after we did the trumpet blasts, I was sitting waiting for it to light up and thinking, 'Boy, everybody is going to really enjoy this. The power is just going to be tremendous and we will be out of our bodies for the next six months; it is going to be so dynamic!'

"The first thing that happened was that the house lights weren't turned out and the thing had started (darkness was essential). I said, 'Already it is not going to come off! Maybe nobody will notice me so I will run back and turn the lights out.' Well, what I did was turn off the cameras, the slide projector; I turned off everything! I didn't even think about it when I almost walked in front of the slide show and cut off the picture. So then I finally got the lights out and didn't think anything about it until somebody came back to say, 'Who was the idiot?' Really I didn't even know until the next day that I was the idiot who had turned everything off! That was my reliving of trying to prevent the crime. (Uriel: This execution was not completed as planned.)

"Then I was supposed to give Uriel the dove. So I sat back there and was waiting with the dove, watching everybody do the controls thinking, 'Boy, these controls! These guys are really reliving. What are they going to do next?' Right before I was going to give Uriel the dove, some people broke the front glass door. That shattered me! All of a sudden I said, 'Oh, brother, we have blown it now!' I didn't even know what the glass was, but I yelled to Danny, 'Close the curtain quickly.' I just wanted the curtain closed thinking, 'We have got to cover this up.' I felt like everything was ruined; we had really made a mistake now and we had to cover the thing. My feelings were, at this point, 'I am going crazy, I am really going to go mad; this is driving me crazy.' I feel like in that past, that actually I did just go screaming, 'Ahhh!' And I then went through the place pulling things out and ripping things apart!"

Uriel: "It so 'happened' that the two costumes Decie made for Calvin and Dan to wear as curtain attendants (which were really executioner's garb, scarcely nothing) were entirely different than all others taking part; all had lots of clothing; the only thing missing was the black hoods; but how thoroughly and detailed does one relive these pasts! It was not told to Stephen just what the program was to be; they simply had him help fix up a stage for entertainment but when he witnessed the murder and found it to be Dalos - he flipped out of consciousness. I believe at that point in the past, Stephen recognized me for the first time."

Calvin: "In this cycle, the whole night I have been reliving being in public relations too. It has been with me strongly because everywhere I have gone, I have met people who have asked about Unarius. I would speak about it in a fairly open way and yet constantly in the back of my mind there would be this thing of, 'Can they understand if I tell them this?' All these questions would come up and I felt good and yet at the same time there was this funny feeling. Then the night of the generator presentation, I was out front here with Dan and we were in this reliving preparation of the assassination, the murder and execution of Uriel in another lifetime. I just felt like Dan and I were the only guys in there; we were dressed totally different than anybody here as we were the curtain openers. I was so insecure and told Stephen who was designing the stage, 'I don't want to go out in front of those people looking like this,' although we both wore these same two, near-bare , silver costumes previously and felt quite proud in them in another capacity or reliving.

"The guilt was strong because I was one who helped set up and announced this presentation and execution and went around saying, 'This is going to be the most gala event that has ever occurred in the universe! Wait until you see what we have done! We are really helping people now.' This is the attitude I know I had in that past life. Although I knew that what I was doing was wrong, yet I would go to these other planets and say, 'Listen, we are really going to help you. We are so advanced we are going to have mercy on you and you are not going to believe how well things are going to go, once you join us.' This is the attitude with which I have talked to people and this is the emotion I have been going through. I am waiting on tables now and have this tremendous fear and anxiety come over me especially during the last few days.

"In going to a table and asking people what they want to eat, I have anxiety come over me and have to grab myself and say, 'Oh, get out of it. Come on, let's just go do it.' But it was that fear and guilt coming in from leading people astray and telling them one thing and knowing that what I was doing was wrong. So, this is my part of the reliving; I had a great deal to do in all aspects of this force field - not just one thing. But to me the main thing was that I was in a public relations type position where I got the message out to people and arranged ways to promote this grand event."

Stanley: "I now remember what I forgot. Almost exactly to the day that Mike Wilson put something in the back of her to support Uriel's spine, I got a tremendous pain. I felt like somebody had hit me in the back with a bat across the spine and it lasted for three days. I have a cyst here at the top of my spine and know that is exactly where the electrodes were implanted. While the whole episode was going on, I felt like my part was successful. This whole advent I had worked on during my whole lifetime was successful!

"It was as if I wasn't concerned if the curtain was working or not. I was only concerned that the generator worked. The part that I did, they messed up - I didn't mess up; what I did was successful! Afterwards, even though it didn't succeed, we all got some kind of commendation, at least I did for all the time I put into it. I tuned in to when somebody works their whole life at a job, they get a golden anniversary watch. I felt that was what I was reliving the other night when Uriel said I had taken the project under my wing. It was like, 'You have worked your whole life on this and now, as your reward, you can live on a pleasure planet,' or some such reward."

Decie: "When Tom was up here last week, I was very much a part of the same type of consciousness - fighting Uriel - saying in my mind he (Tom) was right and to hell with what she said. It took a lot of introspection to admit that as I thought I was being half and half objective with it, but I wasn't because everytime he would say something, I would laugh at what he was saying. That was the way for my lower self to egg him on and keep on bringing those energies through to everybody, to take over so that nobody would believe Uriel anymore, which of course, represents my attitude toward Dalos.

"My part was watching which part of the brain to show them to block off certain impulses and cells and certain electronics which would make her move so that her own body parts couldn't make her move if she was still alive. I have a feeling that he (she) might not have been alive by the time we had taken her all apart. How could she have been alive? I felt that the part of the brain I helped sever was the part that controlled the bodily function of eating, breathing and getting rid of the bodily wastes and I blocked up the whole system. When she was taken apart, I also helped.

"Then when I was in the sewing lab making these big, round collars for the show, it was like stuffing sausages because they were great big, round silver ones that went all around the neck. They just weren't coming out right and I became really frustrated and threw them on the floor and really went into a fit! I tuned in to stuffing some of the body so that we could extract the liquid and then stuff it with something foamy and tissue-like to give a live appearance. I'm sure he was filled with stuffing. I have been having a lot of trouble in that area throughout my entire lifetime with food! My system is very slow in getting rid of the waste products. So I know this is the big guilt of doing all this damage to Dalos.

"When Uriel was up here, standing in the Cosmic Flame, all I could think about was my part in her destruction and what I did to her was all working; that she was functioning okay. Many thoughts went through my mind, 'But why couldn't all the other idiots get their stuff together? I did my part, why can't they do theirs? Don't they know this is for the Orion Empire? This is the last chance to overcome those Forces of Light, if we all fail miserably.' That is when I finally caught my thoughts; I was hurrahing the Orion Empire! With every planet that was being shown on the screen, I was cheering it on; 'Yes, there is another one of ours.' We were taking them one by one, and that total force of destruction came in on me.

"Because of all my guilts, for most of the night I couldn't enjoy what was going on. I was sitting there enjoying judging everything and was totally in my past. Oh, there are so many reactions I had with what I did. She really looked dead with that blue light on her face. Someone put a blue plastic jell over a light just on her face, making her face bluish - deadlike. I actually thought it was a blue powder on her face. I didn't know it was a light and I said, 'Oh god, why did they put that on her? Now she really looks dead! What's the matter with them; don't they know what they are doing?' Then she kept wobbling back and forth and I thought, 'Well, I stabilized the front of her garb, how come?' When I was holding the sun up to her front the other day, I felt I implanted something in her that would hold her up so she would not tilt forward while they had something pushing her back, because they had inserted something in her back, up her spine.

"Then I watched and saw her wobbling sideways and there came the reaction again. I wanted to run up and grab her so she wouldn't fall completely over. I wanted to run to the back and yell at the people controlling the board and it took everything to keep me in my seat and from making a total idiot of myself. (It was not that Decie wanted to save her but that she wanted her part of the work to look good, not destroyed.)

"There is another thing I tuned in to with what Jeff said about the pleasure planet and promising the people that when they died, if they died for the Empire and have given their all, when they came back they could raise hell and have a vacation on the planet of their choice and their life would be free. They wouldn't have to work, they wouldn't have to do anything; it would all be taken care of and the next life would be their big reward!

"All the time I have been working on the costumes I have had this deep, subtle, way-down attitude that I have buried so deeply, that as long as I am working on the costumes, I don't have to look at my past. I am doing fine and I am getting patted on the back. I don't have to look at my past, I am working it out . . . ha! This is what I taught other souls: If they would give their all, they could forget the sinning, forget everything negative that they had done and when they came back, all would be forgiven and they would have a blissful life from there on in! I know it regenerated on this planet with religion, where I also taught religion; I not only taught but lived it and relived it and kept compounding and compounding those energies. I actually believed that as long as I was working in the temple, god would take me to heaven and I would be okay. This is what my actual thoughts have been. It is not the full truth.

"The full truth is that the Unarius principles prove that you have to see what this past is; you have to see what you have done so that you get full benefit in turning those energies around. There is no doing it for us, or that we are going to be taken off and be okay! But I had a big releasement with that yesterday and Mal Var's film also helped. When the lights would come on, at first I thought of the lighted spaceships and the past couple of weeks I have really been feeling excitement within myself about the landing. Ever since I came to Unarius, I always thought I had this great attitude, that 'I can't wait for the Brothers to land. When are they going to land?' Well, deeply seated inside me is this horrible reaction that I actually don't want them to land! They are going to threaten my position (laughingly). What position? I have no position but in the past I guess I did. I was so fearful that my position was going to be taken away and then I realized this was the Brothers' consciousness I see sometimes when I am very transcended.

"The Brothers come in sometimes when I am very transcended. They appear to come in sometimes to my 3rd eye; it is small and then it just beams out and completely encompasses me. When I realized that, the full realization came through! These Lighted ships I had seen at first, represented the last big battle with the Lighted Forces. I was in a craft and they came to round us up, to round me up to help show me what I had been doing and to help me. I committed suicide instead of turning to the Light! This is what that totally represented to me - that I was fearful that I was going to be taken over again. But to say the least, I am glad that the Light always wins out." (Uriel: What a fine admission. No wonder Decie is looking so fine of late!)

Jeff: "It is interesting that I wrote down a list of realizations I wanted to talk about, but I missed one, and through the course of the evening, more information came in so I feel I now have a more relevant understanding of it. When I first came into Unarius, I had a difficult time being around Tom. I was bonkers about the guy and I would do anything for him but I felt uncomfortable and unresolved. I just couldn't ease out. If I went for coffee with him I felt, 'How can this guy waste his time on me?' Well, when he went into his negative cycle with the force field there, I was concerned about him; so very concerned that it finally clicked that my desire was to prove myself to him in various lifetimes because I have hurt or betrayed him in some way.

"This was all kind of jelling in my head and I thought, 'If I get up and talk about it, maybe the rest of it will come in like it so often does.' What clicked in my mind was when Vaughn dressed up as the evil Tyrantus, the Empire leader and he said, 'If you fail the Empire, you fail miserably and you are not welcome here!' I said to myself, 'That's it! Somebody had to take the blame for not being able to get Dalos to submit.' I see myself as giving a very damning testimonial so that Tom took the blame because I wasn't going to stick my neck out. He was up here talking about the time when he came from Pluto; he had this device implanted in the side of his head and I related to it, right away because for the last couple of weeks or so, I have been saying, 'Tom, you think too loud.' Before he says it, I am there with the picture. Then I realized I was monitoring him when he came down here! I was very worried about him at that time, too. Here he was, millions of miles away, he had gone berserk and there was nothing I could do about it, the poor guy.

"But the reason I was so worried was because he was reliving a past that I was involved in and helped with. Well, I said, 'I have to cut these ties and let you go.' Lifetime after lifetime I have been compensating for this any way I can. I would be loyal even if he was wrong, feeling that he was right because in my mind he was Cosmon. That is the attitude I had and it is a good one to get rid of."

Reynolds: "I have taken an active part in this past in a way I am sure I haven't fully realized. I don't know fully the position I had. I assume it was just one of the head technicians involved in working with the engineers and working with the leader of the whole thing. So as a result of this energy, for the first time, as a student, I have never had so much determination to see a thing through. That determination came out sometimes in a very uncontrolled energy. A lot of people felt it as coming from a slave driver - of course I concur with that one hundred percent. I have been a severe taskmaster; a severe person who would be an overseer of that sort for many lifetimes.

"As a result of that guilt, in this present lifetime, I have taken on the personality of a shy, introverted person. Well, that is only to hide that past. I don't want to expose that to anybody including myself. But in this cycle, this energy came into play to begin to work this past out with myself and all the students involved. There were times when I was striving for success; striving to push and get it done because we had a time limit. I felt pressure on me all the time. So, when I really felt myself pushing someone (from the past motivation) I would say to myself, 'Okay, cut it out. Just put the damper on this and damp it down to a balance point.' Of course I would go to the other extreme and would be afraid to tell another what to do. He would ask me what to do and I would say, 'I can't think of anything.' Sometimes I would be blocked off because I didn't want to expose that driving force of, 'You do this and you do that,' relegating authority. I didn't know within myself, in my conscious mind, a balance like I think other people have; like some of you who are more experienced at it and have developed a balance in that way. So this is the first opportunity I have had to let my domineering past out in that particular way. Maybe it won't be the last time.

"Ever since I have been a Unariun I have had this back problem and it has been very close to my spine, but not on the spine. It is more on the left side on the shoulder, on the back. I have had a couple of readings on it and have tried to work with it; I have gone through cycles where it has really given me a problem but when I was working on the Cosmic Flame, I would be wiring the lights into the skirt of the Generator there, working with the light bulbs and fixing them and my back would just be hurting! No matter how much sleep I got or how much rest I was getting, it would just ache. Right when we were removing the costume from Uriel, it hurt the most it ever has. I felt like a dagger was ripping in and twisting. I know that is from the energy of helping to insert the electrodes into her spine and of her spine being taken out and an artificial spine being put back in. All those things were inflicted into my back as well. Many students agree they are all having back trouble in the same area.

"Mal Var said that during the controlling of the bionics, something went wrong and Dalos was fried. Now I didn't hear him say this but a little while after that, I kept seeing Uriel fried - just sizzled to death! I kept saying, 'Did somebody say that?' I didn't want to get up there and say it's my realization if someone has already said it. So I asked him during the break and sure enough he had said it. It just hadn't registered with me. So that is what I kept seeing - that in the very end she went up in flames! I could smell the flesh; I could see all the electronics sizzling right there. I just don't see how anybody could stand to watch it."

Crystal: "Are you saying that the robot went up in flames or Dalos?"

Reynolds: "Dalos."

Tom: "Yes, he was alive."

Reynolds: "Oh, yes, I forgot the audio. I was worried about the audio coming off in a smooth way. I was a little on the paranoid side and went through my check list making sure that all the switches were in the right place, cords connected, etc. Of course, no matter how much you go through a check list, if your past comes in, you are always going to forget. You're going to make a booboo. So when I started the cue, the first thing, the slide projector started to roll its presentation. I had cued my tape for that; it had a sync pulse for Tom's tape presentation, but I had it on fast forward! I hit 'start', and the thing came off! Tom was saying, 'Start it, start it.' I was saying, 'Put on some lights.' So I had to thread the reel-to-reel tape and everyone was waiting for me. Another thing was, that I kept seeing Uriel mouth the words. I had a horrible feeling; it just didn't look right at all. I said, 'Oh, that's terrible!' I was an audio technician who tried to sync up the sound with her lip sync. Apparently it couldn't be done one hundred percent successfully. We talked with other heads of the department to try to come up with the best idea;

to try to pool our ideas.

"So as a result, we were going around with this little control that can be used in the audio department now, that will allow the tape to be speeded up a small percentage - maybe ten percent - if the need exists. Some of the Moderator's first tapes are a little slow and if you speed them up just a little bit they sound normal. That is what I did to the "Infinite Perspectus" series. The original tapes were a little on the slow side, so that's when I put them on a fast reel-to-reel and now it seems normal. That is what my reliving was there.

"I had one last reaction when Bill was manning the camera and wasn't given all the details regarding the program; that was an oversight on my part, maybe. At the first light show with Copland's Fanfare music, we were to have another light show and we were to have soft music so Uriel could speak and I don't think he knew that. He had turned the "F" stop down because he thought that was the end and the house lights were coming on. Well, I kept looking at the monitor waiting for the curtain to open, as I thought it had closed, but the monitor was dark. I thought, 'I can't start the music with the curtain closed.' They kept saying, 'Start the music! Start the music!' I was saying, 'What for? The curtain isn't open yet.' And so I turned around and looked and the curtain was wide open all the time. So I started the music and felt bad about that."

Mike: "I wanted to add something. Uriel was just 'burned up', literally. When I took the huge collar off, all this heat radiated out from her. I wanted to remind you of the fan."

Reynolds: "Oh, yes! We had gone to much trouble to put small fans under her, they were very quiet. We had rigged two of them up and I had a couple of boys make sure they were nice and steady. Stan said to me, 'Let's turn on the fans now,' when Uriel wasn't yet all the way onstage."

Stanley: "She had just gotten in there and I looked and the yellow cord wasn't hooked up like it was supposed to be. I said, 'Get the fans on! Get the fans on!' I wanted to turn them on but . . ."

Reynolds: "Yes, you told me that and I said, 'No, let's wait until she is situated in there', because I didn't want her kicking it because there was no cover on the fan. Well, both of us become so involved that we forgot all about turning the fans on; we were totally blocked off. So I had guilt about not making her as comfortable as I could. I tried to make her comfortable with what we had there with the fans. I tried to get someone to go out there, crawling on his belly and plug up the fans, but nobody would do it. So there is a guilt on my part.

"The spine problem is another thing I forgot to mention. Uriel said she didn't feel like she had a spine when she first sat down. All the time, I was sitting there with the audio, worrying, in the pits, guilt-ridden about her being in pain. I couldn't enjoy the program; I couldn't get an attunement with the Brothers. I said, 'What's wrong with me? Am I really that blocked off?' I couldn't feel anything. No matter how hard I tried to tune in, I just couldn't feel the power coming from it because I was so much into my past. Using those old negative energies of destruction, it just blocked me off from the higher energies. It wasn't until the video program started that I again felt the attunement."

Kathryn: "I was living in Pasadena and came down to one of the meetings and I saw the Cosmic Flame tableau just started; the hoops had been made. I thought, 'I want to work on the Flame,' so I moved back here to El Cajon two weeks later. The skirt to the dress had been started and I was wondering why they had only gotten as far as they had with it, which wasn't very far. Then I started cutting out the holes for the little planets and this was symbolic to me of blowing up a lot of planets.

"As I was sewing the little plastic domes and the waistband around the top, I had to really reach and stretch because the garment was so very large. My back was hurting tremendously, my whole body would become very weak and I would have to sit down for a few moments and say, 'God, I am in so much pain! I don't know what this is but if I keep working on the dress, maybe something will come to me.'

"The only thing I could think of was that I had severed Uriel in one lifetime from her waist up. I was changing the energies by sewing the band around the waist which was symbolic to me of sewing her back together. Well, now that I have heard David's testimonial and other students' about electronics we used and implanted in different parts of her body to make her a robot, this fits in very nicely. I have had many back problems with my spine to the point where I couldn't go to a restaurant and sit through one meal without being in excruciating pain. So when David was talking about these electronics in the spine, these little clues that have been happening came together; my insides began pounding tremendously and the heat came through. So I know there is some healing taking place. Thank you."

Edna: "I really don't know where to begin. It seems as if I am guilty of everything. I would like to thank Uriel and the Brothers for the help they are giving me. I know I am blocked in many ways and it seems that I have so many clues. Tonight while everyone else had been up here talking, I have related to everything. I do know that I turned my back on Dalos. I was in the procession and came in after Dalos and I misrepresented myself in the procession as being the protector of the Sun. I think one of the things I did in that past was to take a life of pleasure and turned my back on the Light. I have been thinking about my lifetime in the last twenty years and realized that my whole life has been one of total pleasure, even though I have always worked.

"Suddenly all the students think I have a lot of money. This always surprised me because I have worked so hard because I couldn't afford not to but I still had a life of pleasure. I have traveled and have done many things. Everything I wanted, I have had. I am now in the United States legally and the last few days I have really wanted to get a job. I hope when I do, I will no longer be psychically living on the pleasure planet. I feel as though I have been 'hiding out' on the pleasure planet, here.

"I would like to thank Uriel from the bottom of my heart for everything she does and for every word she says!"

Patricia: "It is obvious to me that I was working on the brain structures at this time and how to make Uriel's brain work to make certain motor nerves function.

"While watching the film on the brain that Vaughn showed, I got very sick, to the point where I had to get up and couldn't watch it anymore. One word stuck in my mind: basal ganglia, which is the part of the brain which deals with the motor nerves. I was working on the electronic crown headdress, but I couldn't get it right. I knew I was working on the motor nerves. I don't exactly know, (since Dalos was alive), how we manipulated him. I can see how we thought, by stimulating these nerves and cells either chemically or electronically at the base of the brain, we could motivate him.

"One important thing though, I have realized that this is one of the reasons I have had such separation from the students at the time I was working alone. I didn't want to associate with other people because I felt very inferior and because of this, I put on a big facade, that I felt superior and have been going through these two different emotions. I have been typing in the typing room and this was my excuse for not getting involved with the generator and I thought I was being positive.

"Some of the girls got into some workouts and I thought I was just going to be so positive and get this out into the open, when actually the energy I was using was to cause intrigue, strife and dissension. This was pointed out to me by Uriel and I resented it and went into a very obsessive fog for about two weeks until it broke, the night the Cosmic Generator radiated all of those beautiful radiant energies out to us. I realized then, that now was the time to take this opportunity to break through this fog. I had been trying to break through it for four days but I just didn't seem to make it. But with that added boost of energy, I was finally able to feel some energy come back in my life."

Sandlin: "This cycle started for me about three or four months ago; well, it really started when I became involved in the audio department. It is all starting to click. When I first started in the audio department I felt elated. I had the feeling of, 'Now I have it made! I am in the department with David Reynolds who is going to be moving out soon. I will take over the audio department and I won't have to study anymore or work. I had attained this position.' It got to the point where Dave was 'right on'! He was the main man and everything he said was great. It got to the point where I was dependent on the audio department (or that image), and David Reynolds. I started resenting getting up every morning and having to be down at the Center and having to do what David said. I found myself really resenting this dependence on Dave and the audio department for my source of power or esteem, upon which I had completely based this position I had for myself.

"So I was working right along with Dave and it came to the time when we started working on the generator and he was doing the lights. I moved over in that direction. In my mind there was this hierarchy set up, as far as the light bulbs and the technicians go and it was David Reynolds, Stan and I. I had really been going through wanting to be the leader and felt this was my chance. I was in the position where people had to come to me to see what needed to be done and I could exercise this new-found talent of being a leader, or what I conceived to be a leader, which in my own mind and with a lack of intelligence, was telling other people what to do or guiding other people which was what I was doing in a very negative way by saying, 'You are not soldering right; you can't do this.' I was exerting this tremendous ego and it came to the point where I really personalized what I was doing back there. It was my project! And it was pointed out to me by the Brothers through Tom the other day that I wanted everybody to know how great we had built this control panel. I can relate to the dress and how great was the part I had taken in it. The night of the Cosmic Flame, to me, was history in creation.

"I can see that in that past lifetime, we built this robot of Uriel and we got to the point where we spread the news across the planets that Dalos (Uriel) had finally conceded to the Empire and this was a celebration of her joining the Empire. We had the fanfare, the trumpets and the audience. Dalos was wheeled out and he was supposed to make this proclamation of joining the Empire and condoning everything we were doing."

Stanley: "I couldn't get into anything because I was so caught up and concerned as to whether it was going to come off. Everything was out of sync. I was upset with Steve Sandlin for he would point one way and he would turn a planet on the other way and the lips weren't synced to the music and everything was wrong. I was completely upset back here doing the audio. My whole consciousness was on the technicality of why this wasn't going over. It didn't look right; I was hoping Uriel would come out after this, before everybody got up and left, to show everybody she is a normal person or that she was not harmed. (Which really isn't that way.) It was really terrible that I became so critical of the whole thing and that I have been critical for the past three months. Uriel has pointed out to me the ego structure I have built up, on which I have put all my importance. I didn't have a major part in the dressmaking, I wasn't a leader or a taskmaster, but to me it was important that this go off because I put my self-esteem into the dress and for all to go over well.

"So it was a reflection of me and if it failed, I failed. I was more concerned whether this was going to go over for my own self-esteem rather than if the world conceived of Uriel, so it was a personal thing."

Leila: "When I first came to Unarius, the students kept talking about the Brothers and I kept wondering who these Brothers were whom they kept talking about. One morning just as I awakened, I looked up and there was a beautiful Being standing at the foot of my bed. I want to thank the Brothers for dangling the carrot, so to speak, in front of me and helping me to get to this point in my evolution.

"The other day I came to the Center and found out about the list that had been made for us to do to Dalos. I was under the impression that Uriel had made up the list; I became very nauseated when I read it. Later, I thought, 'What will I do to Uriel (Dalos)?' The thought came to make a crown of thorns. I thought, 'Well, this isn't the Jesus cycle. Why make a crown of thorns?' But the next day I went out and made a crown of thorns and then I became involved in sewing and doing different projects and forgot about it. Lately I have been wondering why I had made the crown of thorns. Sunday night I reacted because Uriel's voice was taped and we didn't get to hear her talk. Then I reacted to her hands. I had sprayed the gloves when I worked on them. Well, I realized my reactions were from guilt because I helped make Dalos a robot and I helped project these horrible things to him. Today I watched a movie wherein they used a rod to project severe pain, pleasure or emotions at will. Dials on the rod were used to accomplish this. We used electronics that resulted in the same effects to Dalos."

Reynolds: "Was that "Genesis 2"?"

Leila: "Yes. To get back to the crown of thorns, I realized that ever since I have been in Unarius I have put a crown of thorns on Uriel. I have been so transfixed in my past that I have been projecting to her just like I did then. I projected thoughts like: 'You're too fat. You're too white', things of that nature. I know now this guilt is why I haven't really been able to look at Uriel. Thank you."

Margie: "My part in this Dalos cycle has been involved with the bionics concerning the hands and the fingers because I helped work on the gloves. In testing them, it was like you would add a certain electric stimulation to a certain wire and that would implant it in the hand to react and cause it to move in the way you wanted it to. All this lifetime, I have had an obsession of drawing hands. Everytime I would draw a hand, it would come out deformed. So in experimentation to develop this process, I worked on hands and some of them would turn out to be failures. Also in the experimental development of the pain and the projection of pictures into the person's mind, that would cause him to react with certain reactions. I had a dream about three weeks ago of being at a control panel and in front of me were a multitude of different buttons. On each button was a certain reaction that you could impinge into the person's mind. Typed on the buttons were what the reactions would be. These people were enclosed in a light, clear plastic film and I was pressing buttons and causing all these people different reactions. They would go through all sorts of different pains and anguish and distorted feelings. Then realizing what I was doing inside the dream, I slammed my hands against the control panel saying to myself, 'This can't be but it is, and that is why I have been so robotized and blocked off in communicating with others.' "

Brian: "I was listening to everybody's testimonial and I think it is great that we all have our testimonials to share because there are so many clues to help tune us in. During this whole time of this cycle, I worked a little on the generator here and there. I didn't really conceive of what it was, or care; it was really an ego thing to be a part of it. I was in the Center when Uriel first initiated the cycle, and she was talking to a group of students; I remember feeling as though I wasn't a part of the elite group. I listened to her and ever since then, I have wanted to be part of this elite group that had become all excited as she described the cosmic flame to be built and everything that they were going to do with the generator, and that it would be displayed.

"As we were all upstairs looking at the control panel, I felt really good that day and I didn't feel afraid of Uriel. I didn't have this fear; I did feel good about what I was doing and I felt as though I had a good attunement and I remember saying to Mal Var, 'Wow! This really looks great!' I was telling him how this control panel was really something! I said, 'That will be really great for us for a tunein.' Uriel said, 'What do you mean?

You are not going to be one of the controllers on the ship.' I said, 'Not on the ship but in the reliving when Dalos is going to be tortured.' As I thought back upon it, I now realize that my thoughts were that I really felt good after that happened. When I walked downstairs, I felt I was on top of the world. Now that I think back about it, it really was a chance to show Uriel up in my mind. I didn't fear Uriel. So this just shows you what I have been doing in this cycle, in my mind - one of the things.

"Another of the little definite reasons I knew something was very wrong to do with this cycle was the night of the generator itself. I walked back and I saw Uriel while she was getting ready and she was in this dress and putting the tall pyramid or cone wig on. As soon as I saw her I wanted to get out of there before she saw me, as though I shouldn't have been there like, 'Oh, sorry I saw you and I shouldn't have seen you.' When I look at my thoughts it is really guilt from seeing her defaced in what I saw her as at that moment. I could not face that guilt it was so intense.

"That whole night I wanted her attention really badly but I felt I was constantly deflated by her. My name was one of the names that was on the sheet that Uriel had written of all the things that we were going to do to Dalos and I was one of the persons that should be behind the control board. When I took that part I thought, 'Well, I really am part of this cycle; I have to take it seriously.' I so blocked it off that Uriel had to suggest that I was one of these people so I will have a chance to work it out.

"The night the generator was displayed, I had thoughts when I sat here, 'Is she going to catch on fire? Is there any way? Is there anything that can go wrong; that a spark or something is going to ignite her?' But I had to decide in my mind whether I should ask someone if there was any chance at all, so I could be on guard if something was going to happen. So evidently I was a part of the torturing sequence.

"Saturday, when we put together this song which is the best song I have done, I felt when that came out and we put it all together, we realized we had the whole thing for tomorrow night and we did it in only one night. Once again I felt I was on top of the world; I felt like now I was going to be a success at this festivity! I will be one who is in the spotlight and I can gain some ego-aggrandizement because now whatever it is, has been a success and that makes me feel like I was a politician-type person because everything I have been doing seems so phoney in this cycle.

"Today at work I got an unbelievably sharp pain right next to the spine, in my back. It was like I pulled a muscle with such pain that I couldn't believe its severity. I said to myself that this was some obsession. I didn't know what it was all day and I was going to ask tonight because it was such a dramatic thing. It was like I was fine and all of a sudden, a minute later, there was this sharp pain. I would move a little bit oddly at work because I was walking around trying to hold things and carry them really lightly as I had this sharp pain in my back.

"So with everything that has been said tonight and has been going on, this is a dramatic proof that I was involved in the goings on. I don't know specifically what I was doing but it looks like I was the type of person who wanted to have a leadership part. I had a part in the plans in an outside way but was really seeking my own power and status the whole time. I wasn't really concerned with the Being, Dalos; somehow I must have participated in more of a direct way with Dalos (Uriel), because I am saying right now that I didn't! I know, right as I say it, that I did have a part in Dalos' torture because of the fact that I looked at her before she was ready, backstage, and felt she was being defaced and I didn't want to see her or to face it at all. The fact that I was going to sit at this control board, and of the tension in my psychic, makes me know there is something there that I am not facing; something severe, like a torturer or something that I can't conceive of yet as I am too blocked off to see it.

"I am going to try to open up my mind and be aware of how important this is for me to let go of, realizing too, that all these years that Uriel has come back here to help me. I really want to open up. I know that with study I will find out more details of this life to benefit me in my evolution."

Clark: "I know I have a lot more to realize in this cycle but the one thing I know of is that I worked on the electronics. The electronics went into the brain and I am pretty blocked off to this; I haven't really seen it but I know I was helping in her torture."

Ronald: "A few weeks ago, for the Dalos Cycle, I had a realization of a reliving where I strongly opposed Dalos. I went to great lengths to influence people against Dalos. I was, at the time, fearful that he was going to take over and control all of the planets and this is what I wanted to do! So that was my motivation. I realized just tonight, based on what Uriel had said to me, I must have operated machinery to make electronic robots. I feel, when someone is giving a testimonial on the tortures they were doing at the time of Dalos and I say, 'They are awfully childish,' then I realized that I wanted Dalos' brain so that I could electronically duplicate it and make robots! Thank you."

Dennis: "This cycle represents to me the most negative acts that I have undertaken in my evolution, which of course, is to destroy the physical life of a Higher Being; the second is to represent yourself as one. I have been reliving both of these and there have been several indications. The harmonics that came in as the lifetime in Persia that Stephen and Dan related, I was also involved in that. I feel basically, that anybody who thought he was somebody, who was in a position of power in the government and maybe gave a few orders here and there didn't really have any power. Psychically, I was right in there with this stage, curtain, guillotine and with the construction and concern of how it could have been built better. When Dan and Stephen had the confrontation about the curtain, I was right there with Stephen thinking, 'Dan should just do what he is told.'

"Right after that we were rehearsing one of the dances and I knocked the statue over and cut the head right off. Now, to me, that statue and this entire Center represents or is a part of Uriel and so, a part of the Infinite. So there I was, taking a part in destroying a Higher Being. I was evidently the one wielding the ax. Then the part with the Dalos cycle, the only time I really worked on the Cosmic Flame was a few days before the presentation. I was helping Roberto go over and try out all the planets to make sure they were all hooked up and would light. At one point I was underneath the Cosmic Flame skirt putting the two wires together and at one point he asked me if I needed anymore light in there. I replied, 'No.' Just ten seconds after that, I knocked the light over, the little lamp that was set there to give us light to see and I broke the lamp. So there I was, attempting to destroy the Light!

"But that wasn't the end of it. Wanting to take the light bulb out as I unplugged the lamp, I also disconnected the transformer, so for the longest time we were sitting there not able to get the light to the planets. Of course, the transformer that supplies Light to the earth world planets is Uriel, so I know I had a great part in destroying Dalos in this cycle. I was the one who put the blue plastic on the spotlight; like a good robot,I followed orders. I had a strange feeling about this blue light, because the last time we used a blue light on video, it was for the evil one, Tyrantus, and it was very effective in making him look horrible; but here we were, adding that same frequency to Uriel. As you may have noticed, blue doesn't flatter one's face. Then when I went out in front to view this whole thing and see it, I just had the most violent reaction! Then I smelled something burning; the light up there was burning and smelling and melting the gel; it was not the electronics. So I know that I had a great part in this destruction of Dalos.

"To compound it, the guilt I had was so strong that I went rushing back there and I went up to Tom and said, 'The light is burning, the gel is melting and everything is falling apart and going wrong.' He said, 'Well, everything will be alright.' I ran back and got the ladder to get up to the light. The film was running so everytime you walked by the screen you had to get down on your hands and knees so you would not be seen on the rear screen. I had the ladder while on my hands and knees and crawling across the floor, so fearful! But it shows the guilt I had was so tremendous, now not wanting things to fall apart, not wanting this thing to go up in flames. I was thinking the curtain was going to go up in flames any minute because what I had done to the light was to put foil around it and the gel over that so it could not breathe. This is tremendous, the guilt I have with Uriel, compounded from that lifetime.

"Also the subsequent lifetime there was a time when I played the fake Dalos; I pretended I was He for the following five hundred years. Once he was killed, we wanted to keep up the charade, so in subsequent lifetimes I was made to look like Dalos. As part of working out this cycle, we are doing a scene where that fake Dalos (myself), is being made up electronically and surgically to look like Dalos. We didn't have a picture of me as Dalos, so we did a video tape of me yesterday. It was of me sitting in a chair made up to look like Dalos, spinning around and every once in awhile, moving my head like a robot. I had the strangest feeling the whole time. I felt completely controlled and the controller was the negative force in my own psychic from this negative act! We were taking closeups of the eyes, nose and mouth and I have felt sick and really down ever since filming it yesterday. So there is more and it is just beginning to unravel. Thank you."

Dorothy: "I didn't have any actual part in the building or the sewing or the cutting of the Flame display, but I did go out with Uriel to buy much of the material for it. And all during the time of the delays and the buildup of power with her consciousness on this project, I knew the powers that would be projected would be tremendous. Like Jeff, I thought we would be so transcended that we would not be able to sit in our seats but that wasn't the way with me - I found it didn't turn out that way. I clapped my heart out at the beautiful display of lights and the sky, etc., etc., that Mal Var had put in motion in the background. Then they wheeled Uriel out and I saw her face ashen-grey and her mouth drooped. Actually this was what I was seeing; my whole past came into focus and I had the deepest sorrow and the greatest sadness that I have ever had! Just looking at her the whole time, events of my negative past came forth and I was just overwhelmed with the emotion of it.

"When she and I were talking the next day about the event, Uriel said, 'It was just wonderful how everyone clapped at the motion picture in back of me, but when I came out as the Flame, the applause wasn't there!" She said, 'I was very disappointed, and I felt the students didn't understand what the great flame represented.' I replied, 'I had no reason to applaud. I couldn't applaud; it wasn't there for me. All I saw was this horrible past that you represented with the sadness and all.' She said, 'I felt fine! I wasn't sad!' But she gave us the opportunity to work out in this way.

"I have been involved in the chemistry and the chemicals that were projected in that lifetime, particularly because when the odor came on from the gel smoking, I could hardly stand it. I looked at her and said 'She will never make it! She is so tired-looking and she just won't make it; it is so sickening.'

"I really called on the Brothers for help but I wasn't calling on them for her; I was calling on them to help me get through this horrible thing that I had to face in myself that night."

Gordon: "In that lifetime I had these reactions toward Uriel or Dalos. I got trapped with the machines again thinking that the machine was the ultimate and with the new saw we now have, I thought, 'Well, I don't have to do any more work; the saw is doing most of it. I don't have to have any skills.' It has been proven to me that when I learn how to use this, that it is not the saw that is going to do it, but the person using it. In that lifetime, when Dalos was hooked up to the computer and was supposed to function, I was tuned in to that when Uriel was up here. I had heard before that there were a few things that weren't going to work right, so I felt the whole thing was just going to be a big flop. It was not going right and when she was giving the dissertation before all the lights were on and all the planets in motion, and that blue light was on her face, I thought it was a big failure; this was my reliving in that lifetime; when we failed, I blamed Dalos! I hated him because I thought it was all his fault for not cooperating; for not doing the things that the machine was supposed to make him do!

"So all my hatred was towards Dalos. I started becoming insane with this failure because I wanted to know why it happened. I started to point the finger at all the students, thinking it was their fault. I wanted to take the credit for constructing it before ever getting my hands on any aspect of it. But when the failure happened, then the finger went the other way, 'Of course, I did not have anything to do with it; I took no part in its creation; the other guy did it wrong.' Those were the energies I was using. I had seen the hatred I had toward Uriel, that I had always had towards Uriel and that I had always had towards her (Dalos), in that lifetime, and the thought came to me of all she had done for us this Sunday, and what she had done for us those thousands of years ago!

"She put herself into that position, knowing that which would happen and what she was going to be put through, yet she projected the love and understanding of the Infinite toward us all; toward myself and I didn't meet that love but met it only with hatred. This really hit me the other day. It is something that is deep inside that I feel, though it is hard to explain. I just want to thank Uriel for all the help she has been giving us, no matter what we do to her, because if she ever did give up, we would surely be lost, which reminds me of Michiel's words to Her: "If you ever let me go, the Universe would fall apart.' "

Chapter 6

The *Cosmic Flame* Becomes a Symbol of the Vast Reconstruction Program of the Orion Empire

7/1/79

Uriel: "Well, I wanted to have our unveiling of this beautiful painting by Stephen and Elaine Baldo properly unveiled and if we have enough light, that we can film it, please do it now.

"First let us say that this painting here on my right was done by Elaine Baldo. She presented it to us a time ago. I didn't know that Elaine was such an accomplished artist. We have had it in the Center, using it for different backgrounds and all. I appreciate the pastel colors, but the total applause of this painting goes to Elaine.

"Now, this second one, I understand, she and Stephen and Calvin painted. So, to you three then, goes our greatest appreciation and indeed applause because the way you carried out the colors and the whole meaning shows to me that you students truly can be inspired when you permit it. Oh, yes, do we have the "2001" music for the unveiling, for this we ought to have a "4001"! Let us await the music.

"I'm glad most of you haven't seen this elegant painting; two or three have. I'm glad that most of you have not because it is a very exciting and wonderful portrayal of a spiritual city. (Music is played from "2001"; the audience extends loud applause.) Thank you, dear ones, that is what I call a fine reaction. We're going to get a great deal of benefit and use, joy, and love from this because it truly does appear as our beautiful Shamballa or Unarius. Those spires, spikes and domes really tune me in beautifully. You could look at this for a long time and always find something new. It is truly filled with loveliness and color. I'm sure these three who worked on it had a great joy and blessing. So thank you, thank you, dear ones and I trust there will be many more who will come into this awareness of wanting to do these expressions. Mal Var, did you get a film of that?"

Mal Var:"Oh yes, and that is beautiful!"

Uriel: "Good, I'm glad. Vaughn, we will hear your testimonial, your very important testimonial from this Dalos cycle soon. Of course, Vaughn has been building and building this testimonial for a long, long time. He didn't get to give it last time.

"This is scolding day and what I'm going to say is in the nature of a few reminders. You know, you people need a little prodding and jacking up sometimes and that I place a little crutch under your arm. It seems that if I don't give it, there is no one else who does. Now, you've all been at loose ends working out these horrible past lifetimes. Things have really gotten at loose ends. One thing that comes to mind, first of all, is that I was called down the other night to film at 6:00 and was here, not with bells on (but I was here). I had gone through two sleepless nights oscillating to you all, to help remove all these horrible obsessions and I barely made it! I had just enough energies to do just what was necessary but I was held up for two and a half hours sitting . . . waiting for you all to get ready to film. You see, you have to get the significance here. You are still trying to do just what you did before - oppose, inhibit, delay, resent me and do anything but join in with We Higher Beings. If you begin to recognize these things, they won't happen. Next time I will stay home until you are ready to film me, then I will be down and then I will be sure we have plenty of Power for what you film

"Now, you all seemed to have gotten off the beam or off the notion of study. Why? Let me see how many hands. How many are there of you really, who do sincerely study at least two hours every day? I thought so, I see a few scattered hands.

"For one, when I come in here to the Center, I see Ronald Robbins; he is always studying. Ronald, you are to be commended on your study. I hope you are really conceiving what you read. If we don't gain greater understanding everyday, we are missing the boat. If you were to go to school or go to college and you were to pay these big prices for the schooling, you would study your lessons at home, wouldn't you? (The group says, 'yes'.) You would not just wait for the teacher to teach you and then wait until next time for more teaching; you would study your lesson at home. This is what you must do at Unarius. Every week holds a lesson for you. What I am going to ask you to do that may help you to want to study and because you seem to like to give a report, I am going to ask you to study whatever book you wish, any book whatsoever in the Unarius Library, and whether it is Sunday or Wednesday, each student may give a report. Six or seven minutes ought to be enough time to report because if you take more than that, you won't have enough time for the majority of the group to report. Take any book, but this must be something new that you haven't understood or learned before. Dear ones, you must conceive something new.

"This science must be a growing thing with you. If it isn't, you only revert. Now of the many, many testimonials, really of the thousands of testimonials I have heard on the tapes (and I listen to them on the tape or I read them when they are typed off), I have yet to hear one student say, 'Oh, I have been studying lately and I have gotten so much out of this concept! This concept is just broadening my awareness and my illumination. I am growing with leaps and bounds with that which I have learned today!' Not once do I hear that! All you seem concerned about are your workouts. That's well and good but that's only one small part of it. You must gain some understanding of the science. You have to realize how you implant these negations from your past, how energy works, how you impinge it within your psychic and how it remains there eternally and is constantly oscillating and pouring into your present life day by day, hour by hour existence and that you can't do one thing without this past pouring in. This is what you must begin to be aware of - this past; and don't let it stockpile. Now you can really work out these pasts much more rapidly than you are. I hear you all; you are doubting me, 'Well, I have been at this two weeks!' Not necessarily so, dear ones. If you really obtained an in-depth awareness of what you are doing and why you are doing it, where it is coming from (and you all have had enough clues), then you can work it out in a very brief time. Time is not involved.

"So this is what I set in motion and will expect that you will study whatever book it is. Now I would say, if you are going to take the "Infinite Perspectus" book, stick to that one, and the next time, take up where that leaves off and then report on that.

"We all know that each person has his own level and that the average earth person is not on this Unarius level, but this is why you dear ones are learning what you are - so that we can relate when the time is right, to these people who don't know. The time is right - now; and what you are doing now in these video programs is your way of letting them now know. This is another thing: I would like to have you all take these programs in a much more sincere and serious way. Remember, you are not just play-acting. This gives you an opportunity to relive these pasts, but even more important: This is your way to tell the earth people of this science; this truth. In the enactment of the program, make sure that it is very distinctly spoken. If you need to repeat it a second time, do so because the person who is viewing this is not necessarily aware of this as a reliving, even if the actor says it is such - a reliving. Last night I was viewing the Severinus program and it was beautiful, delightful! The color was great. It was edited beautifully and it was all run off so wonderfully that I really appreciated seeing it again at home on television. I guess while I'm here, I'm oscillating with you all from the input but I do appreciate it in a different way at home.

"This is one little point that I want to make: In the future when you do have or are showing a reliving, it would be well for you to make very distinct and clear just what it represents. I know you mentioned it, Stephen, but the words were expressed so quickly and softly that the average person would scarcely know what you said. Make it specific, as if you are telling someone something for the first time as they are not yet aware of reincarnation. That is my only criticism of that beautiful program.

"So let us enter into these programs wholeheartedly and sincerely with the realization that you are not only acting, you are reenacting; you are reliving from previous lifetimes! This is especially to help the earth people. It is the same and even better than if you would go out and try to tell these people about Unarius. I value these programs that we are making, oh, beyond any word that I can tell you. They are serving a very wonderful purpose. Now that we haven't gotten them on prime television program slots doesn't mean a thing. We will do so in time. It will happen, I promise and which is all a part of your overcoming!

"Now, for these programs that you make of the other planets, I think it is very beneficial that all of you students do attend and watch the filming. I think all of you here should give a little more concern and attention. It isn't only for the few who are taking part, but you should all lend your consciousness. You have no idea how much more beneficial this can be, not only for the person viewing but for the end resulting in the viewing of audiences. You have not yet, dear ones, learned what it is to work together as a unit.

"I wish you could see a film I saw a week ago on Channel 15. It was a documentary movie called "The Kite". It was made in Japan and the film won several important prizes. I'm sure they won them because of their demonstration of unity with all the people. You just wouldn't believe how these little Oriental people had learned to work as a unit. They not only built this gigantic kite that was perhaps twenty-five feet square, but they got it in the air and there were no less than four or five hundred young and older men. They were simply dressed, in little black costumes with a red design.

"They had been doing this contest year after year, lifetime after lifetime. It was stated on the program that they had used this one rope, which was fully an inch and one half in diameter, for over twenty years. They had this program each year in which they sent this gigantic kite into the air and these hundreds of men moved and helped move this kite up or down or left or right to make dives, through the unification of their consciousness. These men, these people, moved like a school of fish; you have never seen anything like it!

"There was nobody to direct, they all felt exactly the same. With one swoop they all went into action and they had some kind of a little hum-beat and it was terrific! It was astounding how those hundreds of people would move and push and pull together. With the many kites, they perhaps numbered in the thousands and covered quite a bit of territory. It wasn't that they just stayed in one little area; it was that eventually one kite downed fifty others because the kite fliers had this unification and all this oneness of consciousness and all this psychic energy pulling in the same direction.

"So if you people, as I say, who are watching when we are doing our programs will lend the actors (those reliving) your psychic energy, your consciousness too, it will help greatly. Live it with them; will you try to do that? (The class says 'yes' in unison.) It could mean a great deal to those participating as actors as well as to the viewer. It will carry an added measure of healing energies.

"Now, our next little project is going to be the "Spirit of Beauty Goddess of Love". It is going to be very simple. It is not going to be a complicated thing like our Generator which took about two months longer than it really should have. It is just the star, like on the back of our number four book. It is to be a big plastic star like that with thirty-three balls, globes, worlds in it. Who wants to take part in making this? Well, that's great; we ought to have it done in one day. That's fine. I think Stephen and Jeff have already drawn it out. Whoever is going to get the plastic, let's see if we can get that before long. Of course the globes (worlds) will need to be lighted. That will be very simple. Then we will need a larger star on the bottom. Now, that doesn't need to be made from plastic. Couldn't that be made from linoleum or wood? It's going to be covered with flowers anyway.

"You know whose talents are going to be called on and he is already overburdened with his projects. Gordon Hook already has more projects than he can do. I just don't want to overburden him. He has a big project coming up there. So someone else ought to make this lower star. It should be that a rather good sized star will be needed for me to stand upon and then the several girls can fill it with flowers."

Mike Wilson: "I'll make it."

Uriel: "Good for you! Another thing we should not have going on when we are filming, we should not have people moving in and out of the door because we get the noise and we get the frustration. How about you, Gordon Jenkins, being the guard at the door? You are so used to being the guard and the door sounds and resounds way in the back. Also, the people in back when they changed costumes, we had to ask for quiet two or three times during the last filming of those in the makeup room. We should realize that anytime there is sound, it will be recorded. If you are back there and must remove things before we are finished, please do it quietly. We have to become much more aware of the camera and the sound when it is running.

"Now of course you know that last Sunday everyone relived instead of receiving us and our Generator in a positive way; everyone relived. You tuned in from your own pasts. It took me just a few minutes to realize and to wonder, 'Why can't they, why don't they and why aren't they applauding? Why aren't they thrilled, happy and excited about all this beauty, lights, etc?' But of course all were reliving in an inphase, negative way!

"Regardless, it's such a wonderful thing that we could bring this symbology into this dimension. The very Confederation (meaning the joining of all these planets) that we attempted before and were so rudely disrupted, and its joining that we are again taking up is the real meaning of this revealing behavior pattern. This is what the Generator is all about. We had planned to film it tonight but I understand one of the lights burned out so we will do it tomorrow evening and anyone who wants to, please come and watch. It will be much better to project energies to people rather than just to cameras. So you are invited; if you wish to, come and observe quietly, about 8:30 p.m., right? That is the gist of my promptings.

"Dear ones, whatever we do say, it is only for the good of you all and only for the good of your progress. (Uriel chokes.) Well, I must say, we do have obsessions galore here. They want to choke me off but I will tell them right now, they don't have a chance. (The audience claps.) Well, that's beautiful. I love your spontaneous reactions! That's wonderful and no, they don't have a chance. They may make me gag and choke and keep me awake at night but they don't have a chance to overcome me. That stopping and blocking as was done so often in past ages is all over with; no more success for the side of the dark forces. We have energies, and power, and Light like we never had before!

"All of you people, each and every one of you dear ones who are overcoming, you are joining in the forces of Light. You are again becoming one of the Light polarities and oscillating in this beautiful, high-frequency way with all of We Brothers. Become more and more aware, dear ones; become more aware of working with us, of being with us in consciousness. This attunement is very important for you to realize and appreciate. When you really begin to live, freed of that negative past, those really horrible negative past lives that influence you so, you wouldn't go back for anything. It's really a complete and totally different way. I am very conscious that every one of you is making effort. Sometimes the effort is done in a little, blind way but you are still making the effort, and as I have said so often, 'Don't fight it. Turn your back on the problem; go pick up a book and get into another consciousness.' Now, I know most of you forget this. When you become negative and when you become bogged down, you forget this but this is so important. You don't oppose, because in opposing, you only lock yourself in all the more tightly. Turn your back on it; just leave it if you want to be rid of the obsessions or whatever it is of your pasts and say, 'Go ahead and I'll just send you out the back way. I'll go over here in the chair and get into a new consciousness!' Read, but read with your mind and you will be surprised how quickly you can change the whole frequency oscillation. It works every time; it works if you put your consciousness into it! You can read with your eyes, but you have to use your mind - your mental function. When you do that, the other and lower has to leave. It has to leave! Okay, dear ones, know truly and know always and sincerely that you do have our constant and never-ending help. When you become bogged down at these times with problems and blocks, tune in to us and that's the good way to tune in. It is pretty hard to do this when you are really bogged down in a very negative way but by picking up a book and tuning in to the text, this is a manner of attunement. We shall be looking for you all on the Inner Planes and during the times that your body sleeps.

"The love of We Brothers flows to each and every one in a never-ending succession of high-frequency Powers and Love and Light. Receive now, this most Infinite Love and Light. Good night, dear ones."

Student: "May we sing, *You Light Up My Life?*"

Uriel: "Well, you certainly may. I think you ought to sing something each time, whatever it is. I think it is a good way to open the session and get you all in a high-spirited way."

Students sing: *You Light Up My Life.*

Uriel: "Very beautiful, dear ones, very beautiful. I'm sorry that we're not going to have the Generator tonight, but we will tomorrow night.

"Come here just a minute, Benno, please. Here is a young man that I am so thankful that he has . . . what are you shutting your eyes for? The camera doesn't like to film closed eyes. Benno had a lot of trouble to get with it, but he has been helping and he's the one who really helped along with Steve Anderson, to make the Generator the success it has become! Thank you, Benno for the change-about. (Everyone claps.) Well, We Brothers love you all very dearly. I am truly proud of all the progress you are making. Now, that's the first step. And now for the second step forward and onward, everyone. Love and good night, everyone! Have a good balance of your meeting."

Chapter 7

July 1, 1979

A Testimonial From Vaughn to the Class

Uriel: "As I was editing the class session meetings, and when coming across this fine article by Vaughn, I was deeply impressed by the great benefits he received as he bared his soul to the other 50 students. Also I became aware that each and every one of these students (would they do likewise in an open confession, so to say), to themselves and the others, gain the similar great amount of soul baring, which could never be measured or related in mere words. It is an experience which only the being, himself, can be fully aware of and too, which cannot be factually conceived of in terms of the total benefits realized.

"The reason is: These benefits are consummated on the inner planes and in the abstract wherein the real self resides. And although the students have heard this lecture by Vaughn, I am sure there is not one who has really heard it in toto and conceived the great and vast message, the complete outpouring and revealing of his own subconscious in the manner that he, Vaughn, literally spread himself upon the floor for others to view! No doubt when these same students read these pages, they will be totally unaware of ever having heard them before.

"So Dotty was kind enough to retype this for me and Ronald, with his ever-ready willingness to duplicate the pages, did so; so we are once again placing these facts before the minds of you students in the sincere wish and hope for you, that you will take these revelations into your own minds in the deepest and most sincere way, and with the knowledge that you, too, could relate similar happenings to the others. For it is true, as you well know, that you have all been psychically tied with this one who has been your teacher, your priest, your controller, your overseer, your overshadowing influence in the past - whatever it may have been that you have been so involved with in the past - yes, eons of time, entailing many thousands of lifetimes. Since this hookup between Vaughn, Tom and Mal Var was made over you, you have not been able to actually think for yourselves! For all that needs be is to relate back to the drastic measures used during the rule of the Empire and by the Emperor Tyrantus in the Orion Nebulae.

"This is to refresh your mind of these horrendous and drastic mind-control measures so expressed by these three controlling egoists, all in the name of and for the sake of extending and inflating their own individual egos; yes, even these three - one over and above the other - have warred between themselves and built up drastic hatreds toward each other. But as the wonderful Science of Unarius teaches, through recognition and acceptance, can healing take place; this, of course, due to the principle of cancellation or reversal of the negative wave forms so impinged when the original negative deed was first set in motion. And what a wonderful thing it is for all persons that this is a valid principle; otherwise people of this and many other worlds would long ago have degenerated themselves back and down, into nothingness!

"So what We Brothers are saying to you in this, Vaughn's soul-baring, is that it should serve you, each one, as a lesson in concept and personal overcoming; for just in these last few weeks since he has so revealed . . . yes, to the world, of his negative and destructive past acts and deeds, he has now raised himself in consciousness to the point where he does now actually care about mankind! His complete and total endeavors now are to help in a selfless way so his victims too can overcome, work out, and begin to shed these chains - stronger than any steel - existing within their own psychic anatomies.

"Student, dear one, take not these words again lightly, as you did when they were first related to you, that they be to you as they were to him - the healing balm; the saving grace that will actually switch your bias and start you truly upon the progressive direction. It is of no benefit and does only regenerate your negative past to continue to cover up, to hide behind, to deny, defy and reject the drastic negative past you all have lived. You must, for your own sakes, if you are to evolve from out these ties of the horrendous measures used by each one, begin to realize that you are not, as so many of you actually believe, a positive force for good. It is not so . . . not so, until you can begin to accept and work with these truths.

"Not until you find it within yourself that you are factually and continuously for We Brothers, the Infinite, the Truth and the Light. And when you do thus experience that you do not, throughout the hours of the day, waver back and forth and find yourself wondering, 'Does she speak truth?', then and then only can you know that you have stepped up on that more progressive pathway. These are decisions that one must make daily, hourly; yes, momentarily, throughout these existing months and years until this more positive consciousness becomes your natural manner of expression. At that time, when you truly are positively biased and oscillating in consciousness with We Brothers, then you can and will become a positive soul, a polarity functioning in the Light, adding your bit to the Infinite and helping those whom you have in the past helped to detour and detain, hypnotize, mesmerize, zap with negative raybeams, etc., etc.

"So you all do have much work to do and much to be gained. The first step begins with acceptance. Now I know you have all heard these words before and the Unarius books are filled with similar statements but the truth of the matter is: you have not yet made them your own. When you do factually begin to realize, to conceive and to admit, as has my dear brother Vaughn herein, then you, too, can begin to experience yourself moving very rapidly forward up that seemingly endless evolutionary climb into the Infinite reaches of Infinity.

"So seek ye within for these answers that Vaughn has so thoroughly portrayed for you, within your own consciousness and when you feel relative or relate to these pronouncements, accept these emotions and reactions and permit this change in bias to take place, whereby these previously accumulated negative wave forms can be switched in their portent and oscillation thus helping to create for you a more peace-filled, happy and regenerative future here and hereafter.

"Know that We Brothers are working with you always, according to your permission for us to come to you, for it does require that you open the door of your consciousness that We may come to you. Of course this cannot be done when these negative blocks towards We Brothers are erected and extended.

"Our Infinite Love, the Radiations from the most High flow to you now from the Infinite Dimensions and Minds."

We, the Unarius Brotherhood,

Uriel

True Confessions Do Indeed Elevate The Soul

July 1, 1979

Vaughn: "I feel I have objectified my part in the Dalos Cycle to a great degree, and I won't say completely because it is never one-hundred percent but to the extent that I now feel very integrated and guiltless. So that indicates to me that I have accepted it in what we call a gut-level. However, I have not particularly talked about it in a public way as there are so many factors that happened. I'm going to have to remove myself from myself and let it all come together in a more balanced way.

"We will start off with the impressions I had at the Sunday "Cosmic Flame" ceremony - the presentation of the great Flame Being that Uriel is now and as she was then, the One Dalos - toward Dalos, the ambassador to the Confederation for the planetary system of the Milky Way Galaxy. This probably is in agreement with many other students who have had a tremendous feeling of fear when the Cosmic Flame was lighted up, that Uriel was expiring as she sat in the dress with all the heat and the various impressions that were received from the light.

"As I looked at Her, and the longer I looked in her direction, I became aware that she was becoming warmer and hotter and that she was in great pain. I had the distinct feeling that she wasn't going to live out the night. I wasn't concerned about her moving her lips with the tape, that didn't bother me. What bothered me was that I felt she was going to expire before we completed our objective wanting to wring from her certain admissions of what I would say was his guilt (I am speaking of the attempt to cause Dalos to appear guilty of duplicity and negation and which He, of course, was not).

"The impression I had, of the fear that she

would be expiring, was truly my fear that Dalos would not last out the ordeal through which we were putting him. We would not get the necessary information that we were attempting to extract from him. I had the desire to again go back behind the scenes to see if I could rectify any problem so that she could carry out the purposes of Dalos' sole objective which was to give us the necessary information on the video so we could have it on the tape to present to the public in that long ago time. Then, when in the present exemplification of that past, when all the lights went on, I saw what seemed to me to be a very positive impression. I was very taken up by the Flame that I saw this Being to be. As you well know, when you see something and when you don't want to know what it really and truly is, from the point of viewing from the negative side, you want to feel that you are seeing something in a positive way. I said to myself, 'Now, I have seen Uriel in a physical way as a Flame Being. There is nothing there but those lights,' but what I was really viewing was her going up in flames! At that moment, I called on the Brothers; I called on the Hierarchy and I called upon Raphiel, Muriel, Michiel and Uriel all at the same time.

"So this means to me that at the moment when she was disintegrated with the faulty electronics, I had a moment of feeling that I had done wrong and a realization that in that split fraction of a second, this was truly an Infinite Being we had put to death. With that recognition, with the horror of it all and with the guilt that came with it, I swiftly moved it behind and rationalized it. I said to myself, 'Well, that was just a fleeting feeling of something!' I didn't want to recognize it. This is true in many instances I have objectified of other lifetimes when I have recognized an Avatar or a great Being as a true Being from a higher dimension.

"Associated with that was the realization that every development in the imprisonment of Dalos indicated that this was no natural, human person; that this individual had some superhuman, supernatural understandings and that we could not bring him down nor could we force this individual to recant, so I realized towards the end that this was no ordinary earth person. This was my great problem here: that I could not accept anyone in a physical vehicle as being other than a physical person.

"It has been my particular position to require any great Advanced Being or Superhuman Being to prove himself or herself in a superhuman way: that is, by some materialization technique they would have to prove they could move some object; that they could come and go and disappear.

"Because of the nature of the Orion Confederation, we had proved to ourselves that we could manipulate the physical worlds; we had various types of electronic instrumentation. We felt very superior because we could draw water out of air, etc. We had this methodology which again was used in other countries in the earth world, particularly by Lemuria and Atlantis. We could prove we had a knowledge of the workings of energy and therefore, why could not this Being prove that he, himself, had superior knowledge? So I could not see in a physical person that which we would call an abstraction of the Infinite.

"The Infinite would have to present itself in more than a normal fashion. That was the cloudiness of my mind, but beyond that point was my own proof that 'I am a superior person!' Therefore, I was attempting it in all ways to prove that I had superiority over anyone who would begin to instill a certain feeling of insecurity within me; that maybe this person really had the goods on me because He really knew more about the true nature of life. He could really show me up! So the instances at the time of Dalos are very far-reaching because they present a continuity of a thousand years or more in which I came from the astral worlds into the physical worlds, took up where I left off and went right back again, trained other people, obviously to take over, then I came back as a child, worked my way up in the family of the then present administrators and the negative leaders of those various worlds.

"Now, as I left the meeting Sunday, I began to introspect and objectify in the quiet hours, other factors of my relationship as my realization of being Tyrantus came in. It was actually happening previous to the filming. This is a series of events that took place and showed me the true vastness of the intent I had. This really brought home to me what Uriel and the Brothers have been trying to get me to recognize and accept in a conscious way - that I was Satan and/or Lucifer. That was a very difficult definition of identity for me to accept, even though impressions, other types of information and emotional feelings would become objectified in a physical way. It would be that this proved to be true in the various things I did. I still had the thought, 'Well, maybe! But how could it be?' and that type of feeling. 'How could I be something like a heinous, murderous, villainous individual like Hitler or Genghis Khan or many others we know of who have crossed the horizons of the earth world and destroyed whatever was in their way or opposing them?'

"I was working here at the Center at night at about one o'clock in the morning. There was only Anderson and Benno here in the back room. They were going through the last phases of checking out the electronics for the electronic display for the Cosmic Flame. Steve came to me and asked me to view what they had finally arrived at. It was a complete success except for one aspect about a whirling sequence of the lights in a spiraling vortex, so I walked in to see. Benno turned the switch and I stood here and saw from this vantage point that it was turned on. As I looked at it the whole left side shorted out on the Cosmic Flame! I asked Benno, 'Did I do that?' He said, 'Oh, no. You didn't do that; it was the foil. It has been shorting out all the electronics.' They thought they had used some inexpensive parts which had caused this but I knew instinctively it was my personal force field!

"Later, the morning of the next day, I mentioned it to Uriel. Actually, they had already told Uriel they had trouble and were rectifying it, which might take a couple of weeks, but they worked all night. Uriel said to me, 'Do you realize that your force field was as negative as that? As your force field presented itself to that great electronic display, which was a representation of the imprisonment of Dalos, it actually shorted out 600 bulbs?' Most people would not accept that understanding, not knowing the true nature of the electronic body. I knew it but it didn't really bother me. It was a knowing but it was still an intellectual knowing. That, along with the viewing of Uriel reenacting the part of Dalos for us and with the other emotional reactions I had built up, the next day as I was standing here looking at the 32 worlds lighted up, I realized the vastness of my relationship to this whole project. I indicted myself with this realization. I saw a vast wall which was electronically hooked up. It represented the same type of wall where they play the war games. They presently have it for interception of any type of instrument that would be sending out a projectile from any other country - as the ICV in Russia. I saw this huge map and upon it were 32 worlds which were shown in electronic display! It was a type of display you see now on the computer boards and on the screen. You could determine exactly what the progress was for each planet, who was there, what they were achieving; the problems, the nature of the country and all the social conditions. Particularly, one could determine what advances were made by the representations of those individuals who were sent to the planets by Tyrantus and by the headquarters of the Orion Empire.

"As I realized it, I knew this had taken thousands of years of effort to come to the point where we would forestall any collaboration with positive people who were representing the true confederation of planets, the Intergalactic Confederation! So I realized that I had meant business; I wasn't fooling around. It wasn't just a 'one time only'; it wasn't simply one battle that would be won. It was a succession of battles leading up to the complete overthrow of anyone who would oppose the people of Orion in their efforts to dominate the entire galaxy! Now that is a pretty vast understanding to have. As I said in a previous testimonial: I didn't really believe in the truth of the stories of Science Fiction. I was always attracted to Science Fiction but I never associated myself as being involved. It took this buildup, over a period of seven years before I could represent myself as being the person who was in charge of all this.

"Now, we have come down to the proof of the pudding. Where am I most of the time when something is going on? I'm never on the scene particularly; I am behind the scene, you see. I'm the man behind the button; behind the person who is pressing the buttons. In other words: From a relationship of persons in control, it starts with President and goes down. Each person is given a particular position in which he has the prerogative to do what he was delegated to do. Now, in this particular position when all of the electronics were being built when the Generator was being made, I withdrew from all of the work of the electronics; I wasn't interested; it wasn't my job as it had been delegated. I knew, of course, when we received this first transmission as to the nature of this great crime in which we here in the nucleus of students of Unarius had been involved, I realized that here was the great opposition and the beginning of the battle that has been shown with Tom Miller and myself. The great opposition we have had has been of the nature of situations where those people who had been delegated authority wanted to expand their authority as negatively oriented earth persons. Power begets power; with power, you feel the need to expand.

"That is the nature of the Infinite - the regeneration in whatever direction. In this relationship, this is how it got to the point, in the progression of many lifetimes, that various people would desire to overthrow the power source and take over. They were successful from time to time. Then, because of the principle of reincarnation, the very person who was overthrown would come back and with great vengeance, turn the tables on the other person.

"This is the sequence of events that has taken place with Tom and me. I have always outwitted him from time to time because I have been more negatively biased and didn't allow my conscience or my higher self to interfere. If it did, I swiftly hid it behind a great deal of rationalization. So, negatively speaking, the opposition that Tom or anyone else has towards another is because they were deflated in their attempt to achieve power. But that is neither here nor there. As far as we were concerned, it is good in the sense - it is a lesson; yet if not understood in its true purpose, it is not a lesson.

"Our karma is with Dalos; with Uriel. It is with the Brotherhood because that is who Dalos represented at that time. I think you would have to go into one of these great headquarters where the center of strategy is developed to see these maps. This is where these great wall maps or these large tables were where they push around the various symbols, as to the nature of the balance of power, which is where you really get the impression.

"Nonetheless, I believe it! It is true because I have always been attracted to this type of film that would be shown either on television or in the movies. I have always been attracted to strategic intelligence or tactical forces. In the Marine Corps, that was my position; I was an aerial photographer attached to Intelligence. I took special courses in polaroid when it was first coming out at M.I.T, where you could distinguish what was being camouflaged. Special types of glasses were just being developed at that time. There is no doubt in my mind, because of my relationship to these various types of techniques, strategy and how to outwit another person; in my whole lifetime here in this earth world, it has been proven to me.

"When I became involved in the business world, my success was through shrewdness. They call it stragegy in military parlance, but in my work I could see what was behind a situation. I could feel the fear and the insecurity of individuals and I would know that this is how I could reach them. In other words, I knew how to touch people in a negative way so they would come forward and be docile. I knew how to use people; I knew how to take advantage of their insecurities rather than to help them overcome them. I reinstated them in a negative way.

"Now this is the nature of a negative leader; this is what we have seen in all the various lives of negative leaders who have lived in our earth world. If you would read their biographies, you would see that they also had that understanding. How else could Hitler have gained mastery over a very enlightened country such as Germany? The Germans were enlightened but they had tremendous insecurity, having been defeated. Their national ego had been deflated. That is what he used to get them on his side; he emphasized national pride. I know these things very well; as a matter of fact, I mentioned this to one of our students here, and when I was visiting our printer in Los Angeles, he offered something along this line to me."

Steve Sandlin:"I would like to say something about what I saw. The last time you sent me for a picture frame I came back and told you that a huge map of that galaxy that I saw was so vivid! When I told it to the class it didn't have any relevancy, but it was exactly as you just described."

Vaughn: "What I am trying to do really, is to build up in your minds certain relationships. Now all of these things happened but they were out of sync at the time. They didn't mean anything. A couple of years ago, I went to Los Angeles to have some of our books bound. While I was there, the owner of the company showed me around. It is very interesting that book-binding has always interested me. So as I was walking out he said, 'How would you like a couple of our books?' He said, 'You can have this one. It is very expensive but it was bound upside down and backwards.' I opened it up and it was "Mein Kampf" by Hitler. I looked at the picture of his face and his eyes and he was alive! I mean these whole astral underworlds are coming out! So I took it with the thought, 'Well, I should have it but I had better hide it away!' I have pushed it out of sight in the office. I read a few words when I returned and it is exactly what I just said. He knew what the problems of his country were and he knew how to bring out this insecurity. But you know, insecurity is all ego; it is a national pride! They felt they had been defeated and were now a second-rate nation. Now Germany was a great country from a positive point of view. All the great composers, scientists and philosophers are from Germany, such as Mozart, Beethoven, Handel, etc. So they had the positive attributes.

Great Spiritual Beings incarnated into Germany to change the nature of the Reformation and the Renaissance. Many of them were German in nationality, but as it is with all Earth people, the Germans, as a nation, developed on the military level.

"Of course, there was that constant relationship between the Light and dark, positive and negative, the desire to bring Light into the world in a national way, through individuals. On the other side there were the military people who were the great military minds. I know of them personally. These are the people who, in that tug of war, finally got the balance. They are the ones who eventually caused the entire nation to arm, not as they had in the past, in little isolated groups, but as a complete national entity to arm themselves to defeat the enemy outside them.

"The enemy was always some individual or leader of some country that was pointing the finger at their weaknesses. Well, this is I; this is Tyrantus. One needs to have a development, and in each of these lives lived is the sum and total and the substance of the nature of his personality at the time he lived in the Orion nebulae of planets. When we are looking at our past from the earth world, you might say that we are looking at the acts we have perpetrated. I am thinking of the most negative lives but singularly they don't have the great vindictiveness they had when they were integrated in a different fashion in Orion because here in the earth world (and I am talking about the time after Atlantis and we can go back to Lemuria too) we didn't have the strength or the knowledge. We really didn't have what we knew and had used in a very strong and dominating fashion when we had lived as various types of leaders in different departments from that long ago Orion time.

"Everyone here, as the Brothers pointed out, headed a department or worked in a department in a leadership fashion. They may not have been head of a department as Tom was head scientist and he was the head scientist during many lifetimes. I, myself, have been a scientist. I know the Brothers have stopped me from incarnating as a scientist as a way of helping me so I would not be able to do those destructive things in a negative fashion. Whatever it is, these realizations though, must come from a different introspective relationship; they must be seen first in terms of the picture which is the intellectual relationship, taken down and then through osmosis, it really and eventually comes into the true consciousness. We could call that 'gut level' where you actually have a tremendous feeling in your solar plexus when the nature of these acts which are presently in our subconscious are beginning to come out of the lower levels. They are now coming into the conscious mind.

"One could not accept this awareness in one fell swoop because it would simply disintegrate the consciousness. I know that if I were told these things years ago, I would have snapped with the realization. On the other hand, I wouldn't have believed it either! It has been a slow development; a slow, steady succession of realizations. A long time ago I used to say, with others, 'When is this going to end?' I mean it is one workout after another; one realization of the nature of our insanity after another. As soon as we have a certain amount of relief, another presentation of information would come in. It would seem we had already seen enough but actually what we were seeing were only certain layers, just like you take the layers off an onion, in a sense. First, you take off the outermost layer which is the thin layer; then you get into the onion peel and you peel off layers and then you come to the kernel. I feel that is what we are into now.

"It will take a lot of introspection and recognition of that inner person who has built up all these layers of ostentatious and unintelligent personality structures. As they are peeled off and you find out what you were doing all along, you feel really good. Then you say to yourself, 'Why this is just a pack of lies. I'm only fooling myself. I thought I was really doing positive things, yet all I was doing was to look out from within the encasement of an artificial personality which I had built so I could use it to fool people!' All of us really know, I'm sure, that we did have a certain amount of spiritual understanding at that time, so we had a conscience. And with that conscience you could say, 'Well, the government of Orion had this vast computer, and it was feeding information. It was writing the newspaper articles. It was distilling the news according to whatever was the need at the time.' It would point out that maybe Planet "X" was developing weaponry to attack us. Therefore, everyone would have to be on guard. So there would be a great military development and of course, it wasn't true. It was a way to get people to work together. These are the things I'm talking about.

"We knew in our true inside consciousness that we were doing these things for our own self ennoblement, as they did in the time of Nazi Germany. None of them can say, 'Well, I didn't know about these great Concentration Camps. I didn't know about the gas ovens.' The ordinary person would say, 'Well, I'm ignorant; it's the government. After all, they are the ones in charge,' but they knew! That's the great guilt complex presently that the German people express today towards the massacre of not only the Jews but of anybody who opposed them! I can say this: that the closest picture we could come to which would be similar to what we did individually and as a group in the many years we lived in this planetary system of Orion, was pictured during the recent World War II.

"These pictures are presently being shown in Germany on television and in the movie films to the children so they will know what actually happened and so it won't happen again. They want the children to know that their fathers and grandfathers were negatively biased and that they were being obsessed by negative leadership so that they, in turn, will never accept anyone who would come around and become what we would call a neo-Nazi in the modern sense and again use the technique of subversion, using the weaknesses of people, for everyone has fears. Everyone is troubled by a lack of personal identity with his spiritual self. This is what is being done today. I don't know how much you have read about what was done or of the propaganda that was used or the military techniques and scientific methods used in their research. I will not go into it at this time, but what they did is closely paralleled with what we did! That's where it was learned. There is no doubt in my mind that the military people of the German Socialist Republic - called in short form, 'Nazis', were people who were friends of ours! We can't say that we are really separated from such crimes against man and spirit.

"I'll tell you a little story and it has never been on tape before. It proves to me another one of these pieces or strings of energy pictures that validate my own involvement. Before World War II, and this was in 1938 to 1940, before the United States was in the war, I picked up a book that was written by a very compassionate, enlightened (you could say) person from the higher worlds. It was about the nature of what was going on in the government of Germany and the true purpose behind it all. He brought out how the Germans considered themselves to be the supermen of the day and that they used Nietzsche as their great counterpart to prove their point. Of course, they distorted Nietzsche's point, as he was talking about the super-ego, which was the superconsciousness. They used the same concept in a national way to point out that the Germans were super because of their literature, their art and their sciences - super to the ordinary earth man. That is what Hitler used again to inflate their egos. You see, it was very clever. I picked up this book and read it and felt I could see; I was there astrally in every scene I read, it was so natural.

"You might not believe this but at night I would converse with Hitler. I would actually be talking with him in a friendly way. Now that was in the astral; that is where my hookup was and that's another proof. In the morning I would come back after I finished the book and remember so distinctly, it surprised me! Somebody called me on the phone, I had just finished the book and I replied in perfect German; I spoke in German but I know no German! There is another coupling of facts that validates my association with the negative forces and which took me up to this very point in time to accept! Thirdly, I remember distinctly thinking mentally (and I don't think I ever expressed this to anyone verbally) about the Nazi extermination of the Jews. It was after the war, of course, but I heard an obsession speaking. It was my lower self saying, 'Serves them right, the Jews. They got what they deserved!' Now, being born into the Jewish faith, you would think that would be the last thing a Jew would say about another Jew, but in this lifetime I was very unconcerned; in fact, I was anti-Semitic. All of this putting-together indicates who was speaking through me and what thoughts I was receiving. They certainly weren't thoughts of a compassionate, concerned individual; yet there was that side of me too but I'm looking back and collecting all of the negative intonations that at that time surprised me. So I put it aside and made light of it. However, now we will get to the true kernel of the corn and that is the onion here.

"This is an admission I am making with respect to my relationship to Uriel. You all know that because the Brothers know everything, they know why we came, of course, and they set it up for us.

They are our teachers. They presented the facts of our past and indicated who and what we were; why, where and how we would overcome. Then we appeared in the womb, grew up as children and were then on our own. However, the Brothers are still helping from the inner, guiding us so that eventually we would find our school and the means by which we would objectify and work out our karma.

"In my position, the only way I could work out my great opposition to the Unarius Brotherhood was to actually live with the leader of the Unarius Brotherhood! I was brought to California in 1965 to meet Dr. Norman, to be in his aura and to have the beginning of the rectification of my lower self. Great powers were directed to me at that time even though I didn't understand it. If you are in the Light in a physical way, greater power can then be brought into the psychic body. That is what was done. I had a series of conferences with him twice more, in 1965, 1966 and 1967. It was known at that time that because of the tremendous sensitivity and vast psychic work of the Moderator that he could see very few people, for if he saw few, then there would be many and it simply was not set up for him to be with the public after He and Uriel met. Their time was totally used in bringing in the work, etc.

"Being what I was and yet not knowing truly, but having a certain awareness of my position, I was invited to live with the great Spiritual Being, Uriel, whom I knew at that time only as Mrs. Norman; however, I did know she was a great Consciousness, right from the beginning. I knew who she was, to a point, but never associated it with a relationship of names or spiritual development. The same with Dr. Norman. I knew he was who he was, but not in the entirety. That awareness was a development. I often wondered when I first came out here, 'Now why am I here? What am I to do? I don't have any particular purpose. There is nothing I know of in a conscious way,' but that was the nature of the beast I was and that I had to be brought completely into a position where this reflection from the true mirror of the Infinite could be shown. That meant that I would see myself. In no other way could I see it because I would see through my feelings and my emotions. I would begin to feel the various insecurities, mentally speaking, in all of the things that developed and begin to realize what I was seeing mentally and feeling was true.

"You see, up until that time, we live in the world and don't pay much attention to these things. We get away, take a vacation or get involved with life and forget. That was my nature to that extent so I was forced to see myself in this way - through this close association with Uriel; to see myself and it was very, very hard for her. Obviously it wasn't easy for me but it was much more difficult for Uriel because she had to constantly project out her powers. Now this is an indication of the great compassion such a Being exemplifies.

"The first life of which I became aware and that I relived, was Pontius Pilate. It took me ten years to recognize and overcome it and then cycles necessitated facing the various other entities representative of negative people on this Earth world up to the very recent time. This will point out how I still had not really overcome, although Napoleon is still not considered to be a negative person, except to some of his enemies in France who were overcome, but to the general population of France and in the history books, Napoleon is considered to have given something of a positive quality to France and to the world - the Napoleonic code and a few other things. Underneath it all, of course, he was a negative entity because he used people to fight his war, and he was a master in logic and strategy in that fashion. So only 200 years ago I could still have the desire to overcome other people through might, through force and by the sword; yet on the other side, he made a quotation, 'In the last instance, the sword will be conquered by the spirit.' He knew this but that was not the direction he took! Knowing it is one thing, but doing something about it is another; therefore, when I came into this lifetime, I was a very sick person. They didn't know whether or not I would live. The mother I was born to was especially chosen to show a great deal of attention and compassion and she gave me the ability to finally walk when I was about two years old. I had difficulty talking and had a great deal of karma in the early years. I apparently didn't want to come in, which has been proven.

"In school I was considered to be mentally deficient. When I was in grade school, from kindergarten to the eighth grade (and I didn't know this until later on when I heard it from my family members) I was very slow in speech and looked half asleep. These again are all the great negative energies overwhelming the positive. The negative energies were so strong, there was hardly any balance, but the Brothers were with me at all times during that period. The whole thing broke through little by little in high school, college and in the service. They were constantly with me but I can tell you of so many instances when I should have been killed, which were the astral forces which didn't want to see me healed - I mean literally dozens and dozens of instances throughout the war.

"Now, I came with all of this to Uriel. I lived in the Light House and lived in her aura and even though I lived in her aura all these years, I have opposed her, mentally speaking. I would project negative energies in many ways, not consciously but I would be opposing her. 'I'll see if I can really get her down; I'll really push her around. I'll do this, I'll try that. I won't give in; I'll listen but I will really not agree.' That's opposition. I came here supposedly to help in the Mission but I came here thinking that Uriel and I were going to move this Mission along!

"When the Moderator passed over, I was going to take his place! Then it would be Uriel and I. I had great plans; I made all sorts of suggestions of what to do. It took me several years to realize that I was a peon. I wasn't in any position to direct; yet the force of my desire to always give directions was so strong! It has only been recently that I have realized that even though the strength was so much a part of me, I would do these things subconsciously anyway. It has been very, very difficult for Uriel and the Brothers but they have succeeded in a small way, then a little more and a little more until they have come to the point where they have been successful and this was the plan that was set up - to succeed in balancing the tremendous lack of equilibrium of this physical world and the astral worlds with which it is associated. So that is my position.

"Dalos represents to me the culmination of my evolution from a million years ago and yet I can't even say that was the beginning, because it was before then. There was a history before that time but the thing is that the negative descent of Lucifer crystallized with the Orion Confederation! When Lucifer was thrown out of Heaven, literally speaking, he was told, 'Look, if you cannot go along and be a Brother and work with us instead of trying to make your point and make suggestions as to how we ought to continue with our efforts to help other astral realms, you are then indicating that you are against the Light.' It was a very serious affair, he being one of the Brothers, which was during the time they gave him many opportunities to recognize what his attunement was with the lower worlds, which he once was a part of before he was Lucifer, because it was rise and fall, rise and fall. This was the third time - and out! That's why it is so serious. This is what Uriel and the Brothers have been so impressing upon my mind ever since I came out to California to help with the Mission of Unarius - that I pleaded on my hands and knees for an opportunity to come back.

"I was told that it would be very difficult; it would be a superhuman effort but if I would try, they would help that much more. The trying was done. Of course, it could never have been achieved if Uriel's aura was not surrounding me in a physical way, as well as it is, in a spiritual sense. So the crystallization of the most negative development where there was a plan (as I pointed out) an electronic chart, showing how we were going to overcome these various planets was devised.

"Now, the 32 planets are only a small part. I didn't indicate it but those are the 32 planets that were in danger of leaving our fold at that time. You see, we were a part of it but we had begun to recede. We were then turning people around to recognize another leader. The Confederation was headed by the Spiritual Forces and Dalos recognized this. He was coming to the main headquarters planet to see Tyrantus to try to change this tremendous desire to overwhelm and destroy the plan that would change the vibration, the oscillation or the frequency of the entire galaxy so that this galaxy would eventually become an astral world because this was (is) a physical world.

"Now the worlds have to change. Let's go back to worlds; go back to individuals: How can we change our social structure? Each individual must be a link in the chain. This is the same with these worlds. This is a vast awareness to have, but this is the nature of the Unarius Brotherhood. It is a vast organization of Spiritual Beings, vast beyond the concept of any physical mind. They have charge of not just one galaxy but hundreds of thousands of galaxies that they are administering to. The Milky Way galaxy was in line to become and develop into a higher octave and eventually to extend the nature of the Infinite in a positive direction. So the Infinite expands out by the nature of each infinite spark. These infinite sparks are always overshadowed by one or two, or thousands who are chosen to overshadow various worlds. We were the counter force, you see. We were the opposite force. At that time the balance had not been made yet so that is the great challenge we face at this time; we are looking at our true life source now. And if we do not accept the fact that at one time we were students of Spiritual Beings before and after Dalos, if we do not recognize that we knew that which was wrong and yet chose to go with certain parties and lend our hands to eventually overthrow the Confederation and not work with the Spiritual Brothers, to overthrow the Orion Kingdom, then we will really be straddling the fence, you see!

"Think, as I have, of the great horrors that took place at the time of Hitler, a time closest to us. Hitler lived the life of the Genghis Khan who did horrible things in his overthrow of the various countries; not only China, but he came all the way down to India and to what are now various parts of Iran, Afghanistan, Syria, Lebanon and which was the whole Asiatic continent and almost into Germany. He became the force that he was through destroying the physical bodies of these people. He never stopped short of killing men, women and children in the most atrocious ways. Now, Hitler did it in the more scientific way and hidden away from the public; he took the people on railway cars to outlying camps and there, in the isolation of these camps and in other buildings, destroyed these human beings.

"We saw and we have been told and we have recognized that we have used human beings as decoration, as fodder and fertilizer. We stripped their skin. We took their various parts of the physical anatomy - the bones and tendons. We have used the brain to study it, just as we showed on the film. Hitler's scientists did the same thing. They took the skin, as you know, and made lamp shades of it. They used the individuals as guinea pigs and tried injecting various types of chemicals to see what would happen, so you can't say that the Hitler regime is not a regeneration of Orion. It is! I am the personification of a person who stands very close to Hitler in the past. I considered him to be what we call a brother spirit, in a sense, although I am not proud of it. And horrible as it may seem, I did at that time have a feeling, a rapport.

"Now, I have had a tremendous healing. As you know, my son is here. I have karma with him in my decision to achieve things over and beyond the concern of other people and that goes for the family of people. I deserted him in many lifetimes to achieve my own purposes for self. I would always rationalize it away, 'Well, he's not important. The government is more important; the nation is more important,' and so on. Larry was the son of Napoleon when Napoleon was removed to Saint Helena, and his son was taken away from him to live in Austria. Napoleon was not allowed to receive any letters from him. He was nine years old at the time.

"I left Larry when he was nine years old in this present lifetime. The purpose was to help overcome and work out my own karma, this time for a positive reason. At that time, the reason was because I was more concerned with achieving national ego. It goes back, back and back, but the only reason I am bringing this up is this: I have the opportunity now to rectify all of the negative feelings of abandonment which developed into resentment at this time, by having him with me for a few months. That could only come to pass because I had overcome various aspects of my lower self whereby now I could be able to work with him in a positive way, because he is the other side of me. He is the positive polarity and I was the negative polarity. I went this way; he stayed firm. That doesn't mean that he doesn't have any negative life experiences but they were not life, after life, after life whereas he would go out of the physical world and into the spiritual worlds. When I would leave the physical world, I would go into the lower astral, strengthening my desire to come back again. So it is a wonderful awareness that there is something that is on the horizon for all of us. I never thought this would happen; I never thought I could overcome or I could break down or that I could become integrated with myself and with those whom I had damaged. That goes, not only for the members of my family, but for all kinds of people - whomever I meet.

"The healing that I personify is a healing up to this point. It isn't complete but it is a healing where my guilt has been changed. That means that in not having the great guilt, the negative energy wave forms have been turned around. It means there is hope in store for all people but before it will be, you have to live in the pits to recognize that there is where you have been - mentally speaking. You also have to know that it is a special trial course. It's like the racing car that is being proven on the proving ground. It is going round and round the track to see if the chassis is good and strong; to check the motor and every aspect. This is the proving ground we are in, but there is a purpose. There is a rainbow at the end. Whatever it is that you feel is the lacking substance in your life, in yourself, that lacking substance can be brought to fulfillment and achievement in only one way; that is: to dig out the dirt that is covering up that true, inner, individual self.

"This Mission is successful because of the fact that we are facing our past and we are able to look at the representation of what we have done to the greatest Spiritual Being who has ever visited this world aside from Dr. Norman. Uriel is on the same par. What she has done to help us progress and become healed is just inconceivable. I don't believe I, personally, will be able to understand completely because we are still being fooled by the physical nature of this body you look at. Until we have removed all of the guilt from us, we won't see the true mental being that is there. That is what I feel I saw for a flash second when all the lights were on during the presentation and I saw this Flame Being. I see how it is that even at that time, I could have repented and changed my whole evolution if, at that time, I had realized what I had done. 'You have just killed a great Spirit; you have killed the Infinite - a part of the Infinite; at least, the physical part. You have destroyed it in its attempt to help others. That is all He came for.' If I had realized that, you see, then everything would have been different. Naturally if Jesus had not hung on the cross, we would not have the great guilt but with the nature of our great negative past, that was in the works you might say, although that again is another story.

"So friends, I say to you that what you see in the physical is nothing . . . . nothing! This physical world absolutely represents a passageway and that is all. If you become stuck in a passageway like a cork gets pushed into the bottle, you are going to have a very long time pulling out that cork! You need a special mechanism to dislodge it and get it back to the passageway and that is just what we are doing. But that bottle with the wine and all of the pleasure factors represents only one way to recognize the congruity of the Infinite. That congruity is positive and negative and is understood by recognition of the principle that lies therein. Don't become pulled into the schoolroom itself and believe this is the nature of life, because it is only a passing speck of time.

"Well, I have taken up a great deal of time here

but I'm sure it was necessary. If anybody has any questions he wants to ask to bring out any aspects here that were clouded, I'll remain for a few minutes."

Mike: "What was your position at the time of Dalos?"

Vaughn: "I was the Emperor Tyrantus. I put into effect all of these plans. Alright, I didn't go into that. I am responsible for all of the repercussions that have taken place. I sat with the committee of all of the department heads and, of course, all of these people were the specialists. I was the generalist who formulated the plans. It was my suggestion that we do something to obstruct Dalos from intervening with the plan, of changing our desire to overwhelm the galaxies. We were on our way until he visited us and presented us with the awful part and our future that would take place. He tried every way to indicate what our future would be and how our spiritual selves would be damaged to such a great extent. He practically read our future lives for us. I had already decided in advance, before he came, and my ear was deaf because I didn't believe this person was someone who had any greater or more special understanding than I did!

"In other words, it was a show of strength - my ego to what I thought was his ego because I didn't believe this individual represented a higher source than I did. That is how insane I was. I would think deeply at nighttime what methods could be used. I would have in my hand all the various people who would give me ideas and suggestions; not that they would necessarily work, but as to whether they were feasible. It was a group plan; working out of a method. The method was finally developed, building this massive building completely saturated with electronics. Whatever we have now is nothing in computer science to that which we had developed then and actually put to use. We had even invented other techniques. Maybe at a later time we can go into it.

"So, I was the emperor; I represented that individual. I was helped by the Brothers to replay or enact the part for our film. That was a setup from the Inner Worlds where it was decided. All of this is what we had decided to do and what they had planned for us to do, so we would represent our true negative self at that time. Now I didn't believe I was the emperor even though I had that flash. It began coming in and out. It took me . . . I don't know how many hours to recognize who I was by dressing in that uniform and costume, putting on the makeup and looking at myself and saying, 'Yes, there I am. This is I.' That was Tyrantus speaking to Uriel when she was the Peacock Princess and sitting on the throne. That was Tyrantus then who should have pleaded for help and admitted his wrongs at that time. That was Dalos, not the Peacock Princess, really, although it was a woman in a woman's body but that was really Dalos. The idea there was the changing around. You see how the Brothers work? That wasn't planned, by the way. That came five minutes before she came in. There I was with Dalos, the real Dalos, with the mentality of Dalos. That is my subconscious. There was Dalos and there was Tyrantus! Let me say that was not a slip of the tongue. Obviously I wanted to be Dalos. I wanted to be whatever that individual represented as a great intelligence, a great strength and stamina and the complete composure. That was a turning around of the past at that time when Dalos came to Tyrantus and pleaded with him with logic and reason.

"Now it was the opposite. This is the positive representation where I had finally come around and repented. Everything I said there, and I haven't the slightest idea what I said, by the way, in conscious mind but I know from my inner self that I recognized that I was a spiritual being and that I was not representing spirit. I was regenerating the devils in hell and I recognized all of the consequences that had followed for a million years. That is what that represented. You see, there are so many facets to the true healing that took place. Before the presentation of the Cosmic Flame, I had already recognized who I was. I really knew that was I. When I was sitting in that chair for the film "Severus", that was Tyrantus speaking with complete, free rein, and I was listening. The other self was listening to it, but right after it was finished, I couldn't get out of that costume fast enough, pull off the baldheaded wig, tear that moustache and beard off. Oh, I just did not like it! I had already begun to recognize the nature of what it is that I was. There you have it! Anyone else?"

Roberto: "The decision to get rid of Dalos and terminate his life: Was that the Emperor's involvement or was that just the thoughts of some of the department heads?"

Vaughn: "Well, that has to come from you people. I don't feel that I wanted to be or was in charge of the scientific sector and what was going on. Our desire there was to develop ways and means to perpetrate the belief that Dalos was on our side so he would represent that which we represented. We were trying to change the nature and the logic of what he was bringing to represent us as being the negative people. We were using him. We changed all the connotations and then we would get Dalos to speak in our favor so that he would represent one of the leaders of Orion. There are certain students here who know of the true relationship of the electronic guilt that took place. We had some kind of a beginning of a reliving with that aspect. Now it is coming back to help in this way.

"An emperor is not an emperor by word alone. By being a leader, you have to know what is going on. You have got to have your finger in the pie. It is the same with all governments. That is what spies are for. Every department where there are specialists had spies of the enemy. They were hand-picked people who would report to him what was going on. You saw that wall map representing what was going on in other planets. Well, at the same time, these representatives would report back to these other planets electronically. They would come through spaceship travel and through other media, voice contact, etc., but on the planet itself and in our own planet there was constant contact with the various other people who would tell me what was going on in the department. I knew what was going on so what I found out, of course, was that there was a plan that this department head, and that department head were coming together. They were going to begin the formation of a coalition to oppose me. Those people were dealt with - destroyed! I would replace people from time to time if things didn't work well and I would tell them to get out and they would be replaced by someone else. Only those who were loyal would stay. They would have to prove it through the various types of electronic surveillance, until such time when they could fool the emperor somehow by subverting the electronic systems because he knew what was going on.

"Now in the office I use in back of the Center, there are several mirrors on the front of the door. This was done during our workout in Lemuria by me. I realized at that time that those mirrors were electronically controlled in that past. I had something like a spectrogram in the office where an individual would be in front of the door and the entire nature of that person would be shown. A readout would be given; I would know what he was there for, who he was, what his past was and that which he had on his mind at the time, then he would be admitted. If he was admitted, he would prove the nature of his guilt by not indicating that he was speaking the truth; that would validate it. This is one way the negative leaders kept their hold on the public. I won't say I knew everything that was going on because there is such a thing as negative forces fighting among themselves. Thieves fall out, and this was the same thing from time to time. This, I believe, is the situation that occurred here; something happened where the scientific sector was attempting to develop the superiority and to achieve some power within the power structure itself."

Patricia: "Before the Cosmic Flame presentations there was the jubilee and the parade, etc. Did we have a parade or something?"

Vaughn: "Well, that was gone into. That parade was when Dalos came to the planet. He was invited as a dignitary and received in the way of one of his stature, presented with gifts, with entertainment and honored with flowers. I served at that time but did not know that at that time I was the ambassador for the Earth world. I still didn't know what the Ambassador represented, and was the one Tyrantus! That is how we are helped. If I had known, I wouldn't have touched these realizations with a ten-foot pole. I wouldn't have wanted to know about it, but it was all set up by various types of ceremonies and special effects. We thought we were doing something really nice but that was the representation of Dalos, you see, in his space costume sitting on the throne!"

Patricia: "Was it the same parade they had with the trumpeteers and everything?"

Vaughn: "The same thing. It still has a negative connotation with the trumpeteers. It was all for the same purpose - to help those people who were involved to recognize their relationship. That was for the public, by the way. It is being televised and that was being used so we could show it to the people - how we were treating this great Being; yet that great Being was doing what we wanted him to, being a robot for our purposes. Some of you have testimonials that will go on for several sessions into the future. No, it isn't easy to do these things; you have got to suffer! The physical body doesn't mean a thing. If you are hot, thirsty and all that, get a drink but while the power is here, built up as it is, we should not dissipate it. Uriel knows that I had to set the pace here because whether I said certain words or didn't, you will see a reflection of yourself in some way.

"Now let me say one word of caution to you. Just because I have indicated to you that I have had great inner recognition and I have greater peace of mind than I have ever had, and a greater dedication to the truth and my higher self is much more integrated, now I can go out in the world and see the Brothers and feel them, no matter where I am and know which is the right position. I give little attention to the physical world; my mind is on the continuity of the consciousness from the higher self, from that end coming down. My consciousness thinks here, but that doesn't mean I am finished, that I have completed my karma and I have made it. I haven't. That lower self is there; it is a computer and it is manipulating the life of the present time, because you couldn't live in this life if you didn't have it but it's not in-phase to that great extent.

"Now I still get negative thoughts, but I know what they are. They don't throw me. I know immediately, you see and I change it, and it works! I used to carry these thoughts; now I recognize them and they go out. It is just like I swept it under the carpet but they will always crawl out again. So that is the caution. You have to learn and know the principle. That's the nature of our overcoming. You can't just keep on having workouts. You know what will happen? You'll become a nothing; you will feel like a nothing. You will be a person who has no direction because you will feel so whittled down that you won't feel you can stand and face anybody. That is not the purpose; the purpose is to fill the gap in with new information to gain strength in the psychic. Then when that spiritual self is reaching out from the physical, you feel ten feet tall. You are not aware of yourself; you're not concerned what people think of you because you are speaking from another dimension. Obviously when you are receiving from the past and feeling such torturous guilt for the crimes you have committed and all the various things that go along with it - insecurity, lack of confidence and all that negation, obviously you can't have that integration, but you know this is what we are all learning.

"I am surprised to hear that when Uriel asked the question, 'How many people study two hours a day', . . . well, how many people who are taking the classes are studying? You know when you go to the classes you are studying at least, but it is still not on your own. What she meant really is: How many people are studying actually like they are going to medical school and actually studying a particular lesson and conceiving a chapter in the morning, in the evening, whatever - not just here, listening to a tape.

"The Brothers purposely set up these Unarius courses in the colleges and that will be the future for the world. It will be a hard course to take. You have got to actually prove that you are involved! The course means that you must start analyzing your negative self. You must have the positive part, too, which is the part relating the principles. So, there are two sides to this equation. What she said is true for Wednesday and Sunday classes both. Instead of volunteering, whoever is moderating the meeting is going to indiscriminately point to a person and say, 'Okay, give us a review of what you conceived from the chapter that you read during the last few days.' In the class it is a definite must that you have your assignment, where you actually come with a report of the inner realization of what is going on in your subconscious. The rest of the evening and whatever time it takes, you can say what you want.

"Uriel likes to see Dave wearing that Conclave ribbon. It is a very positive thing. So, if you have one I think you should wear it to all the meetings. I have a few extra in the office. At least wear your Unarius pin at all times."

A word from Uriel: "Dear students, this fine admittance Vaughn has given should be given much thought. He is indeed freeing himself of much heavy karma in these confessions to self, for that is who he is really admitting to and it is a real benefit. In this awareness and revealing, he is striving to help free you all, too, as you were, each one and countless other Earth people, all connected with him - controlled to do his bidding. So his testimonial could well be a similar one for each student.

"So I suggest that you take not lightly that which he states in these testimonial times but take it within yourselves and benefit therefrom, knowing your karma is entwined with him and the more he can release, so can you, too, as well . . . with the awareness!

"Remember, We Brothers are with you always, helping in this freeing process!"

Chapter 8

July 1, 1979

Student Testimonials

Dave Keymas: "I have had the realization the last couple of days that I spent many lifetimes on these pleasure worlds we created in the Orion Empire. It has regenerated down onto this Earth world, back into Lemuria in the sex cults and the orgies that went on and back into the times of Babylonia and the Middle East with the harem. I have had these harems and spent entire lifetimes in nothing but pleasure, drugs, sex and power plays on people. This is important for me because it is such a strong part of my consciousness. I have had this really shown to me in a direct way lately. I know it is very true because every time I think about that I start getting a tremendous headache. I know these headaches are all the negative forces that I'm hooked up with from these lifetimes. They don't want to be released; they don't want to be exposed. Even though I am speaking now, the pain is getting worse which is even more proof. I want to discharge these energies and be rid of it so I can move on.

"I went to Disneyland a couple of days ago and all I could see was a gigantic pleasure center, sophisticated in the ways of providing pleasure for as many people as possible. I was so impressed with the efficiency and the way they would move people in and out of these rides. They would have a car emptied and a car filled practically before you could snap your fingers. Signs are all over saying, 'Disneyland, The Happiest Place in the World', or some similar description. This is what these worlds were. 'Come to our Pleasure Planet, the Happiest Planet in the Galaxy'. This is what I relived with Macinus when we enacted the Idonus presentation. I was thoroughly indulging in the pleasures of the flesh, or drugs and hallucinogenics that made me feel like I was something when I was nothing . . . nobody. I needed drugs. I needed many women around me. My life was complete as far as I was concerned but it was shallow. I can't continue to believe that. I have got to be rid of that and replace it with something, or as Vaughn said, we will feel like a nobody if we continue to live the way we have in the past. So, hopefully, with this realization I can break out of this lethargic, apathetic consciousness I have been in and get some good studying in. Thank you."

Vaughn: "I want to make an announcement here. Uriel wants you all to know that Frank Garlock has again given a fine, sizable check to the Mission! (Everyone applauds.) I can say about Frank, that right from the beginning when he made the contact with Unarius in 1974, he has always known that he was going to play a part. He has never given any thought to the nature of his help because he knew he was going to play a part although at the time he still wasn't aware of the karma with the various people. His higher self was inspiring him; he always paid attention to that self, his true self, and when he gives it isn't done to be noticed or to receive praise. It is simply because he knows that this is what he has set up to do - all a part of his preconditioning.

"This is the best way he knows how, at this time, to help the Mission because the funds are used for the video and audio and go into the nature of our various presentations so Frank is actually seeing a part of himself expressed, in a sense, in the purchase of equipment and so forth. I wish I could do likewise. I do really wish I could come up with a large sum but one has to do this completely selflessly because Uriel never likes us to use the word owe. You don't owe the Infinite because the Infinite is infinite. You owe it to yourself to recognize how you can overcome your karma. So, thank you, Frank from us all!"

Frank: "I want to point out that each and every one here and those home study students are doing what they can too. I wish to state that no matter what it is that one does, that when he does it in the right order to help the Light, it certainly is appreciated. We have all been prepared in different ways to do things at different times. It is a great joy to be able to do this. It's not the number of dollars that counts, it is the energy and the feeling that goes into it that really counts and the return is multi-multi-fold.

"I always remember the story about the mustard seed from years ago. Well, it is working here at Unarius very well. A point I wanted to bring out is that whether one does it in a physical way, whether one does it in building property or he works on a particular project or if out earning a few dollars or bring-it in that way, I think it is just great what everyone here is doing. So I don't want to take any special credit here at this time because of the number of the dollars. There is a lot of sincere, hard work going on and Vaughn, you have been in there putting in. If there's any comparison going on, which we would not want to do, there is a lot of sincere effort and time and a lot of caring for us. I want to extend my personal thanks to Vaughn for what he has done. He speaks of dollars but he has done a great deal in labor and time with us too! I really appreciate the help I have been getting.

"Two other things helped tune me in to this particular cycle of the reliving of this time of Dalos in this liaison position between the Emperor and the other sub-leaders. David Keymas came over with a carnation. At first I thought it was for Barbara but it turned out that it was really for me. I put it on and it helped me to kind of keynote this position of moving about and being in a very responsible position during that evening. Also the head of security was always checking in with me to see if everything was alright, further to set me up. So these are some things I wanted to relate.

"Also, I want to really make it a point here

because Uriel reminded us that we need to speak up and let be heard, what these Unarius Principles are doing for us. To me it is not totally learning about my negative self because there are some good moments that go on too. The principles help us to realize those. I had a question asked of me this morning, if I knew who a person was in a certain lifetime. I decided I wasn't going to mention it unless that person could figure it out for herself. I feel like I can make a statement now that might be interesting to prove the lighter side, a little joy of life here, the peace of mind of learning what your negative is and turning it around, but also that there are some nice things that have occurred in our past. I would like to introduce at this time, since this person knows who she is, the one who is very dedicated and very caring and very dependable, as one Mrs. Fouche, of that time. Would that person like to stand now so you will know officially? I thought you would be interested. I will sign off at this point. Thank you all very much."

Vaughn: "As everything is always from your past, that was from the past, Frank, introducing your wife whom you, in that past, had just married. Mrs. Fouche was a noblewoman, you see. You, Fouche, were a commoner and Bonaparte's second-in-command. Of course you had been given titles by Napoleon and others and you married an aristocrat, you see. You were given these titles and that is what you were doing: You were speaking to an audience at a reception. It is really interesting for here and now you are totally reliving!"

Kathy: "I'd like to thank the Brothers for the dissertation on the Unholy Six Cycle that they brought through Vaughn. It has helped me a lot. When I do read it with my mind's eye, I get a great deal out of it. I have had many realizations and healings from the recognitions of these pasts that I have lived in this Unholy Six time.

"One thing I do know is that when Dalos came to

that planet, I was there to help receive Dalos. Also, I was a surgeon. I worked on these little electrodes that we implanted into Dalos' body when he was in this little room or force field.

"The night that Uriel was in the planet dress, I saw roaring flames come from her. When I heard the sparks, I smelled a scent and thought I knew that he was being burned. But I knew that the Brothers could take care of any malfunctions if they were to occur in the electronics so I wasn't concerned about Uriel in this current presentation.

"I was sent to a pleasure planet. I was sent before the fake Dalos came. He was Macinus from the planet Idonus. Tonight Vaughn mentioned that some people who were turning against the Emperor would be snuffed out or sent on. As he was speaking, his eyes were focused on my eyes. Just before that moment I had felt, 'Well, I would like to hear some more of what my part was in this past, I am ready and willing to listen to it.' So I feel this is one of the reasons I was sent before the fake Dalos was implanted in this particular pleasure planet. Another reason why I believe I was sent was that I was a plastic surgeon. What we had done was to implant these little devices into the physical bodies in many areas that would give the person pleasure from the exterior. This would make the body climax in all areas. I find this to be true in this lifetime, that no matter where I focus or in the actual contact of physical relationship, if someone were to pay attention to my elbow or to a little piece of flesh on my arm, this would incur a releasement with my body. This has been answered now, or the reason why this does occur to the exterior of the physical. This is a big part I had played. I do, and am recognizing at all times now when this occurs because I want to be able to have this past work its way out and have this energy changed so I can indulge in a relationship of a more natural order and not be hooked up to this electronic pleasure mechanism. Also this inhibiting device for Dalos was hooked up to this computer that was as large as a small planet. So there is a lot of karma that I have incurred in this area."

Vaughn: "I have to speak again. This shows you just what I said: how it is never over with. Roberto, in regard to your question you asked as to who was responsible for Dalos going up in smoke? I was responsible! That was the plan. After we had finished with Dalos and done everything we could possibly do to misrepresent his true purpose, it was my suggestion that we get rid of Dalos in a way which would suit our purpose. We would show that Dalos, who was now an accessory to our plans and our purposes, was really a representative from a higher world. We would dematerialize a person in a flashing, blind light. This was to show here what was going on. This was shown to the masses because we had to have some way as an 'out' after Dalos was killed. Of course, the Emperor, as crafty and shrewd as he was, used every technique and scheme possible. So here he said to himself, 'This is a way for us to get an advantage on both sides of this situation. We will represent Dalos as having visited our world and as having partaken in everything we have to offer. After all, we treated him in the highest, most royal fashion. Then we will put the ceremony on tape and show how we had been visited by a higher Being from the great beyond. We would indicate how he returned, with all these lights that went on and I am sure it was much more magnificent in terms of its power. They would see this physical person; then they would see all the lights going all at once. The lights would die down and there wouldn't be anyone there! That body was completely dematerialized by the tremendous heat; it was electrocuted, to all intents and purposes. That is what I saw when I was viewing it. I said, 'There's a real Flame Being!'

"Actually, I saw that from the other side of my higher self. I could see that this was a Flame Being really, just as when the final act was played and Jesus was hung on the cross - I had the same recognition! I had killed a higher Being, a spiritual Being! That was why I later committed suicide. This is the same understanding I have. I didn't realize it until a few moments ago as I was going around with it."

Dennis: "Vaughn, I feel that I was set up as the fake Dalos to appear on the other planets to show essentially, the godlike quality of Dalos. Then we planted a fake Dalos now representing Orion. Isn't this factual?"

Vaughn: "That's right."

Dennis: "With the electronic devices implanted so that a little flash would show there was mental psycho-kinetical power that would come through?"

Vaughn: "Sure. The fake Dalos would represent only the true, the good and the beautiful, yes."

Dennis: "Which would be our representation - Orion!"

Vaughn: "Right. That's what we were trying to represent - who we really were; that we weren't really bad eggs."

Michael: "Was there only one fake Dalos?"

Vaughn: "At this present time, I don't know of any other. There could have been several for other planets!"

Dan: "I thought there were duplicates sent out to other worlds."

Vaughn: "Well, these worlds were hundreds and thousands of light years separated from each other. Some of them had no contact with each other at the time. I'm bringing that to the point here that it is true. That is what the Brothers are saying - that the Emperor didn't stoop to but one single counterpart. He would see to it that all these planets would have the subterfuge of another leader. Of course the intention of Dalos was to seed these planets. That was the purpose: to bring to these leaders who had been planted there the true nature of what they were purposely set up to do.

"Uriel mentioned this a long time back when the group met and when we were receiving these books on the 33 World Planets. I remember when we were in the Lemurian Cycle, at the beginning of it, she said to me, 'Vaughn, you've been following me around to each of these planets, haven't you?' We were in a certain recognition. I could understand that I was, but I didn't know then what I had done. So, the answer is yes. It had not yet been revealed."

Patricia: "Friday I went to Disneyland and it was quite an experience for me, being that I had visitors from out of state I thought I should be entertaining. I was very tense because I didn't want to go in the first place. I was thinking, 'How am I going to get out of it?' I was afraid they would think there was something wrong or weird. I was reliving that I was on a starship. They were the inspection crew coming. I had to show them that everything was in working condition because the whole day I relived this, up until a point of death. The day when I came back, I wrote down everything, only it was in a different time period.

"An interesting thing was that my brother has been involved in the Unholy Six Cycle ever since I can remember. He had a book called *Blood Hype*. It was about a drug that when taken one would become instantly addicted. It would also cause death or disfigurement as one used it. I became very sick to my stomach when I saw that. I saw a movie that was so horrible, I saw it twice. In it I saw a parasite that was very corrosive. This person put it in drinks, and it would just burn the person up from inside. I know that I have used these drugs very negatively in the past. You'll have to excuse me for I am just trembling. I can't get in front of people and talk."

Dan: "Referring back to Kathryn's testimonial about the pleasure planet, I had a dream the other night in which I feel I was on this pleasure planet. A woman came up to me and wanted to have her sex organ enlarged. I performed the operation, so I know I worked on that planet in one or more lifetimes.

"Referring to what happened last Sunday, with the presentation of the Cosmic Flame, I was reliving like everyone else. Uriel asked me after the whole affair was over if I felt the power. I told her that I had been aware of it for the last two or three days, but that night was the ultimate in my evolution, though in a roundabout way, I was avoiding answering the question truthfully. The honest truth was that I wasn't really feeling the power, I was reliving. I was too negative. I was inphase with this ego part of myself that was trying to tell Dalos I was fine and I didn't need his help!

"Last Sunday Tom said that the flame we created did not work. It was not successful; the electronics were not working. Those controls hooked up to her mouth were not in sync with the audio that was coming out. The planets weren't lighting up at the proper time when she was speaking of them. There was some talk that something went wrong, that possibly there was some sabotage. A day or two later three or four of us together found ourselves at a table listening to Vaughn and talking about this. I realized after listening to all of this that I was one of his spies during this time. We were trying to determine if there was some kind of sabotage that took place. Did it come out the way we wanted it to or was this just some kind of technical error?

"I remembered someone mentioned that Margie had gotten up and said she had a dream that she was on this console and all of a sudden became frustrated and just hit all the buttons at once. She said, 'I can't accept this.' Someone said it appears that Margie is the one who caused this malfunction or problem or whatever it was, yet I still didn't feel that was true. I feel I had some kind of inside information in that lifetime. I knew the one who was responsible for this sabotage but I didn't tell the Emperor for my own reasons. This is what I feel. I have my own ideas but I am not going to say who it was or whom I think it was because I feel it is up to that individual to come to his own realization. I do feel I should say what I feel here about my own reliving with that.

"The other part I wanted to speak about was last Sunday after Uriel left with several students, I tuned in to this resentment toward her which I have felt very strongly in this cycle. It has come to my attention as to why I have this resentment. This resentment really started back when I was the leader of a physical planet. One of the commanders of the Orion Empire came to me and solicited my support and promised to make me a god or a leader if I went along with him. I have come into the actual emotions and feelings that I have had during this lifetime as leader of this planet. It has come about because I have taken this job at Farrell's.

"What happened is that I have seen two diverse sides of people. I see how we as a group are constantly reliving and seemingly bickering among ourselves and being negative much of the time. Then I saw what seemed to me like a more positive side of people's nature on this outside; how these people, even though they react to each other would not hold onto it. I mean it was for only a few minutes and then they are positive again. They are going on. What I was doing was reliving the way I then looked at the people on my planet. I was supposedly there to help them. They were always having problems and needing help. I was sick of being around people who were always wanting me to help them. This was the way I was looking at things at this particular time when this commander from Orion approached me. So this is why I went over to that side. I was tired of helping other people whom I felt were inferior to me and whom I felt were sick. I thought I was somebody who was above them. I saw an opportunity to serve myself and do what I wanted to do. So I did, I did what I wanted to do.

"In doing this, I went against my own inner self but I wouldn't recognize this. I went against Uriel who appeared to me psychically during that lifetime to tell me that I would make a mistake if I left. So I have personalized this resentment upon Uriel who represents the Light, who is the Light, who is the Brotherhood. The resentment is not really against her, it is against the Infinite. It is against the positive forces. It is resentment that is the result of this guilt constantly pressuring in on me for having made the wrong choice and not accepting it, not recognizing this as fact; not being willing to say, 'I am wrong. I did the wrong thing. I am doing the wrong thing.' My ego is so large that there is this force in me that just does not want to admit it is wrong and negative. This is what I have been reliving for the last month or so, I would say."

Crystal: "My last testimonial stated that I have been reliving that which I have created. These are deeds and acts that I have done to other people. I know that I have been reaping the fruits of being a very negative force and the repercussions from this. I have to realize this; I know that the type of leader I was, was one who extinguished anyone whom I thought was getting ahead of me. I have been reliving this in my consciousness and it is a horrible thing. When I see any of the students doing something progressive I get this horrible feeling of doom that comes over me as if I want to stop them. Now I have to recognize that it is my own negative force that I have built up that actually wants to stop this person from doing anything that is progressive. In so doing, I have stifled my own creativity in this cycle. I haven't had any inspiration. One of the main reasons for that is that I have been realizing many things and I haven't been refilling those holes or those gaps in my psychic by study. I have been tearing myself apart, beating myself, realization upon realization but as Vaughn stated, it is important to refill and to have something to replace that energy. That is something that I've got to start doing and that is studying or there is not going to be anything left of me.

"Now the major thing here is I feel blown apart several times a day and my whole psychic seems to blow apart just like it is on a vibrator. It is a very negative feeling. If I understood correctly, Vaughn just said he instituted a plan to have Dalos extinguished. Is that it? Was it set up that a panel of people would go in and set the fuses so they would get rid of Dalos? Well, when we were sitting at this little panel yesterday, which was a reliving, that was what I said. I said, 'Vaughn, do you think it could be that we eventually got tired or disgusted that we couldn't break down Dalos, so we said, 'Let's just get rid of him and get this thing over with and initiate Phase III, or whatever was our plan?' I'm the one who brought that up.

"It is really quite a lesson that what we have done to others we do to ourselves. We have to relive it. Today Uriel spent a lot of time with me because she knows the seriousness of my condition. I just want to thank her so much. I have been in such great opposition, reliving all of this satanic, terrible, negative force against her as Dalos, that it wants to kill this part of me that wants to ascend. I'm not feeling sorry for myself but I am really accepting this. It feels good because I haven't wanted to accept this. It is like things come in and I will immediately put up a wall. From not accepting, I have just buried myself. (She cries.) All this time this was going on with Dalos, I was studying him very closely because I wanted to be like him. I wanted to figure out the superhuman strength and higher energies that he had and yet I wanted to destroy it too.

"All down through the ages, I have followed Uriel and have tried to impersonate her and other Spiritual Beings; so much so, that there is little of me left. I am just big obsessions. So, this cycle is going to be by far the most beneficial one. It has got to be because like it was said earlier, if I don't face this horrible past, it will pull me under. I want to thank Uriel from every bit of my soul for the help she is giving me."

Dennis: "The few testimonials tonight have given me even more of an insight now to all of the intrigue that went on during this time and the complexities of the vast collection of insane minds. I know I was one of the most insane because I also was very responsible for the death of Dalos. Being set up as the impersonator of Dalos, I can see very clearly the personality that I was who would push very strongly to get rid of the real thing because it just shows up the real lack of integration I had. If we had gotten rid of Dalos, then there would be only one, and that would have been me.

"So I know I influenced Roberto by offering him the pleasure planet and all its glory if he would do certain things to the electronic equipment. We relived this when I helped him check out the Cosmic Flame. As I said in an earlier testimonial, the smell and the burning plastic came from a light covering that I had placed up there. So in the last 24 hours I have just begun to really accept that I killed Dalos and that my own insecurities and ego were responsible for the death of this Higher Being!

"I have had the heaviest feeling ever since that night of the Cosmic Flame because I haven't been able to accept this. Uriel had pointed this out to me in one way or another. It is a very vile brew that I have refused to even look at because the smell and the taste are so bitter. It's me, it's a part of me; I have just uncovered the pot. I haven't even begun to look at it but I know the reality of it now. I know the truth of this negative self. So I am going to change it now."

Vaughn: "We have a lot to think about. You can be sure that until you, personally, objectify your part, you won't have any relief. There's obviously more to it but I think we are coming to the crest and we are going downhill, at least in terms of our participation. Next Wednesday and next Sunday and the next Sunday after that, I believe we will have come to the culmination of this particular period. You know it will come around again. Take advantage. So let's go home and study. Good night."

Chapter 9

Reattuning to the Very Worst and Most Negative Pasts Ever Lived by Unariun Students

Then Comes the Freeing

June 13, 1979

Channel: "It seems kind of early to film; it is still daylight. I guess that's the best time to start which is when it's Light, when it's dark! Tonight we are going to talk about the Unholy Six Cycle. There is a list Uriel made up that various participants, in attempting to work out, could do some derogatory things to her to work out this very horrible, negative past; the feelings of resentment and jealousy and negativity toward her can be worked out en masse or in a large part, if we enter into this with the attitude of working out, and we want to rid ourselves of these feelings incurred in lifetimes such as the one when she was Dalos.

"It was originally expounded by Uriel that we would physically do these various things to her to reenact the Dalos living, while she was in the force field, as the reliving stated, and that with the accomplishment of this in a physical way, we can have our realizations and breakdown, and we can damp the energies, the negative forces that we have had against the Light.

"However, in the time of Dalos, and the time of the Empire taking over the various planets with which they were involved, they were very highly technical people. They had machines doing much of their work for them; even torturing people. They had arrived at the point where they could institute energies into the person so he would hallucinate, and he could believe he was in another environment and actually, that physical things were done to him. There would be physical distortions take place with the body; there could be other people doing physical things to this person and he would see it as a real thing, such as throwing things at him, putting spiders and snakes in the vicinity to create fear; having sex in front of him, orgies, etc. These are some of the things Uriel thought about that we could be doing (not actually) like tearing her clothes off, tearing off her hair and those tying her up, etc. This could all be generated by this advanced, sophisticated computer device, and still all these mental images and even all these outside, environmental conditionings for that person, when actually nothing physical happened to Dalos at all. But it was all projected within the mind and this is the way we are going to depict it.

"We are all going to have our relivings and we are all going to do these physical things to her, whoever becomes involved, and yet we are going to place Uriel in a force field in a white room besides that. We will have a computer in front; Mal Var is making it and we will put that right in front where the people are pushing the buttons. The people can change; whoever is pushing the button, will be reenacting the physical act, in the physical. Yet in the video, we will zoom in on Uriel's face and put her out of focus and cut to what is in her mind, this being done when the button is pressed, like making her face distorted and all that. The person who is doing that reliving will actually be the one who has pressed that button. He's out there pressing that button; we zoom in on Uriel's face and we slip-focus, cut to your actually doing that, then we cut back, slip-focus to Uriel, then put her in focus and zoom out to a clear, clean immaculate room. Nothing has happened to her physically, yet she is in agony because she has experienced all that in her mind. So we all have our workout and we show how it really was. Dalos was not physically touched and yet all these things were projected into his mind through the form of electronics."

Steve Anderson: "I have lost track of what the

reason was for keeping him captive."

Answer: "It was to try to get him to follow in the Empire, to join them. They knew that if they could get him on their side (our side), then there would be a powerful force there. Any positive person who turns negative (if he can be induced to turn negative), then they have access to a tremendous force there. The negative force knew this and they were acting out of those negatively-logical reasons, believing that if they could get him, then they would become strengthened. He was saying, 'No', and they said, 'Okay, then we will distort your mind so that you will relent and we will have power over you!' Even with all these mind distorting things happening to him, nothing physically happened to Dalos. Yet, to the person to whom it's happening, it is very real because the mind is everything! And if they threw images, if the computer threw hallucinogenic images such as people doing all kinds of weird acts in front of him that could be a holographic projection, but nothing really physical. And yet the mind can be told to receive images and the one actually being manipulated, actually thinks it's being physically done to him."

Bill: "Do you think that it was almost like telepathy?"

Answer: "No, it was done by computer; it wasn't done by mental means."

Bill: "How do you mean?"

Answer: "The computer actually did it to the brain. You see, everybody who comes into the physical world has fears and phobias throughout their entire incarnation. They have dealt with snakes and spiders; were killed by men, and doing all of these things. They are full of fear and so the subconscious mind can be interfered with by higher states of consciousness. There is a static condition, a broken line of communication taking place, then the fears and the phobias can be interjected and the person has that reaction. Now we are not saying we cut the psychic anatomy, the Higher Self of Uriel off, because she would die; anybody would die if the Higher Self were cut off. But there was this interference pattern that created subconscious, involuntary reactions to negatively stimulate fears, phobias and various reactions. I hope everybody understands that and that it is quite clear."

Stanley: "The other night you were talking about *Star Trek* and how almost every single episode is from the Unholy Six Cycle. Well, about two weeks ago there was one, just exactly like what you just said - torturing One such as Dalos and exactly like it to a 'T'!"

Answer: "Yes, because it says that Dalos was fastened down and put in a force field and all this. Actually in thinking about all this, visualize if you will the high technology necessary to put up such a force field; to make computers, to make spaceships and then actually to cause one to physically get down and grovel; to throw cans and bottles and physically do these things, tear clothes off and do lewd dancing and all this and joking in front of him.

"All of that would be messy, for one thing and it doesn't relate logically to me, even in a negative way, why that should be done when it could be computerized and be accomplished in a very clean, hands-off manner, showing that you do have power over that person in being able to manipulate his mind processes and also to actually turn it off at will to show that the person has no control over what reactions are real and what are not real rather than to get in there and boo and hiss the person, and all. That is really degrading and debasing our own selves even more negatively. And we had all the electronics at our disposal to create all those images, so wouldn't it be natural and normal to use it to show that we have more power over another person?

"And even then, Dalos (Uriel), did not succumb or relent, because from the very start, to the end she knew this was all mind-induced, anyway. This was what they did not understand: that even though they could induce these images in his mind, when he came out of it, he knew what had happened; that it was all an illusion; it was all fake. And this is why they couldn't break him down because even playing on his fears and frustrations, he was wise enough and had such high contact that these things had no effect. This is why the people who were pushing those buttons got so angry: they were doing their utmost, to no avail. Anybody who was put in that machine before, broke down just like that (a snap of the finger), into a raving lunatic screaming, 'Let me out of here! Let me out of here!' So I think that doing these physical acts is the only way we can overcome these things because we have no way to project our minds or imagine these things as real to work them out. We need to actually get down to throwing these things or doing these things to her in the physical so we can work these things out on this basis. This is why I think this is set up for us to do this in a physical way because in this way we really tune in to the medicine that we were throwing at other people, especially him. These are the energies we are using right now - today! Even though we might not today physically pick up a can and throw it at her, we surely would all like to sometimes because she deflates our ego and she points up things that are so true that we can't get out of it and we hate her for this in the present time.

"Yes, we do know that in the end, the result will be that she is helping us although there is still that momentary gut reaction of 'Why are you exposing me? How dare you not look up to me! How dare you think any less of me than I think of myself, I who am at god level?' And Uriel is constantly letting us know our weaknesses so that we can overcome these weaknesses which are why we react, and until a person learns and begins to understand how energy operates and how we are creatures of habit and how we are electronic robots until we start to use these energies selectively, we are not our own man doing our own things. So we are learning to overcome these pasts by actually getting into them! That's why these dramas are so tremendously important to any one of us. We should not shirk our chance to get into our workouts, but we should realize this is one way we can actually, physically reincarnate these experiences into this dimension. Then with having them before us and facing them, we generate an out-of-phase field so that these healing energies from the Higher Brothers via the channelship of Uriel are instrumentally bringing these frequencies into this dimension. They cancel out the very pernicious effect of these negative happenings and are cancelling the negative force that we have instilled into them; the power that we have given them in our everyday life as reactionary values.

"We are actually physically going through these things as part of the play, and in doing the play, we can actually show how it is all mind-induced by performing the effect as we described to you earlier. That will be very effective."

Brian: "Do you want to actually do physical things, people touching her physically?"

Answer: "Yes. This is the only way to work these things out and this is what we said: We cannot work them out by just mentally saying, 'Yes, I threw a rotten egg at her.' You have to act out that energy right now because in the past when you pushed that button, you had the computer do all that and you just stood back but mentally, you were projecting the image of that negative act or else you couldn't have pushed that button. Now we have those energies, those hate frequencies and torturing frequencies in our psychics, radiating out. And the only way we can get rid of them now is to physically go about and do it because we don't have a computer as an intermediary right now, and we won't be freed of those past activities until we work those energies out; we won't have any kind of motivation to use instrumentation positively in our lives. We would not use them, we would abuse them! Are there anymore thoughts on that?"

Christine: "Is that why we like to keep a distance from her now, physical distance?"

Answer: "I think that is individual. I don't have that 'wanting to keep a physical distance from her.' If one does have that feeling, I would say that it is very likely there is tremendous guilt, and not wanting to be face to face with that torturous energy that we used in the past, yes."

Stephen: "I would like to bring up one other thing. Uriel said that anybody who wanted to be made up or wear a mask at the time so they can get it out and not be recognized, it will be okay. In other words, anyone who would feel inhibited to be, themselves, seen doing it."

Answer: "Well, that's fine if Uriel says so. If you want my opinion, you wouldn't get the benefit if you are hiding behind something. You might as well not do it at all."

Gordon Jenkins: "But the makeup you would have on like the makeup they had in Valneza is much like a natural mask without putting a mask on."

Answer: "That is just my opinion. I wouldn't want to wear a mask if I were going to do it."

Jeff:"Maybe some people will get tuned in just by seeing the mask."

Answer: "Well, it's just a personal opinion. Uriel brought it up for a reason, so we will all go around with it. Personally, I wouldn't want to wear one." (Maybe she was planning on de-masking you at the time you were harming her!)

Dan: "How realistic are these things going to be?"

Answer: "We will see about that. We want to make it as realistic as we can and yet not dangerous. We'll work out something one way or another. Maybe we can have an expression of Uriel's face and then have a jar with a tarantula, a real one, and we will get a closeup of the real tarantula. Maybe we will go to the zoo and get some snake shots and things like that."

Stephen: "She mentioned a couple of things in that regard today. There is a way to make styrofoam rocks and bricks that will look real. We will make watered-down food coloring stuff for ink and paints to throw on her so that it doesn't make a permanent stain. Then she will be made up later to look like the after effects of laser beams, etc."

David Reynolds: "My gosh!"

Answer: "Yes, this is really heavy stuff. Uriel has a whole list and she thinks we should do more than that, even. You can speak to Dorothy about that."

Dorothy: "I have been in the midst of this workout and I am sick half the time. I could tell you all the things that have happened to her through me and my frequencies. I have been sick and nauseated and I would like to get it on tape."

Answer: "Okay, when we have the testimonial time, you will be the first one up, if you want."

Channel: "There are a couple of concepts that we want to get into tonight. There have been rumors or thoughts coursing through your minds that Uriel has been aware of. I think it is important tonight that we go into this concept a little more deeply so that we can get a better understanding. Some of us might already have this understanding, but it is imperative that we do: that the Infinite is the Infinite Intelligence and there is no other outside force than the Infinite. The Infinite is the sum and total, even the substance of all things. Therefore, there is not an evil force in the universe and also a good force in the universe, as being two separate forces. The Infinite is just that - one force; a unit unto itself, contained within itself. And it does not have two forces within itself, but it is one force or power. The Infinite is the force; is the substance; and the totality of all things. It is the power; it is the life-sustaining systems.

"In other words, it is everything that you can conceive or otherwise; even beyond the horizon of your own consciousness into the higher Celestial Worlds and above and beyond that is the Infinite Mind. It is also, in the same respect, the lower astral worlds and the Earth plane. It is all Infinite Intelligence; therefore, 'What makes evil evaluations, and again, good evaluations?' It is how that power, that substance or that force is used! Simply put: It is how this universal force is used by anybody who determines its subdivision into evil and good, or how it is positively or negatively biased!

"So let us not say that the evil forces are fighting the good forces and the one who wins out will annihilate the other force. What you are saying is that the Infinite is a two-headed dragon, trying to devour itself. But it is not! The Infinite is simply an impersonal force and therefore does not put hooks and nails within itself to rip and tear itself apart. No matter what stamp of approval or disapproval you put on any energy in the Infinite, it is still the Infinite. It is really not outside yourself because all that you are, is attached to the sum and total of this unified force. Therefore, in condemning or avowing or disavowing any information that comes to you is the property of the Infinite and whatever you might think about it, or however you might interpret it will not change it.

"I am sure that you all think this is a fairly simple, straightforward concept, and yet as long as there are judgments, prejudices and evaluations of good and evil, and superior and inferior feelings, then we have not accomplished that one vital conceptual link to the Infinite. That is: There is no separation of any form of life from another. Even from the most devilish and heinous or fiendish criminal in the Infinite to the most Hierarchial or highly Celestial Being, they are both all a part of the Infinite; they are the Infinite, using the force of this Infinite in two diametrically or oppositely poled situations and thereby placing the frequency in totally different dimensions.

"This is how the infinite force, the universal force keeps the integration of itself moving forward and dynamically, and that is the frequency relationship principle. But just because you are oscillating or moving dynamically in the higher dimensional perspectives or dimensions, does not make you more a part of the Infinite nor are you separate from another Infinite which is over here, in another direction frequency-wise, in a devilish dimension. To understand that vital principle is to arrive at a nonjudgmental position in life, allowing all forms of life to seek their own level because it is all part of the Infinite, no matter how you, in your own frame of conception divide it and subdivide it. In other words, all the opinions in the universe do not change the face of the Infinite. It is one force, and how it is used determines what dimension or consciousness you are in.

"Surely the Brothers want us to understand this concept because it is simply stated, yet so importantly underlying many of our other premises of concept, and if misunderstood can bring about many misconceptions and false conceptions.

"There is another concept that we wanted to get into with the Unholy Six conditions. Speaking of the chronological aspect of how this Empire started, how it progressed and declined, is important for those souls involved to understand so they can place themselves in situations that are viable experiences in their day-to-day, present lifetime. The mish-mash of the concepts that some of us come up with sometimes, evaluating them as realities and experiences that actually happened, sometimes do not have the quality of real intelligence. For you see, in experiencing these things, these negative, horrendous ties, the year-after-year, lifetime-after-lifetime knuckling under of our own negative forces to negative leadership and the obsessions involved, has caused us to become so submerged in this miasma of negative plasma that we developed in a manner such that caused us to be unable to determine what was real and what was not real. So in subjecting this Higher Being to these torturous events in the mind, and we, being the perpetrators, we subsequently had these distortions reenacted in our own state of consciousness and these horrible, obsessive forces came into our consciousness and whispered all kinds of horrendous deeds that we are accusing ourselves of physically doing.

"Much of the time it is simply to keep us submerged in that state of consciousness that would keep us confused and keep us from really realizing what we really did and what we wanted to do and what we incurred as consequences because we did certain acts. So, living in the Unholy Six period, established for us a hoard of obsessive factors and you cannot cast these energies unto the Ones of these higher echelons of dimensional understanding such as Dalos and not expect them to come back equally torturous to us in a future time. And thus it was, that for lifetime-after-lifetime, our own lives became so obsessive that we did not know whether these energies were real to us or not real.

"So it is in this lifetime. If you get all of this chaotic energy into your mind, and all of these devious and devouring types of energies, do not just accept them on surface value, but evaluate them with true, realistic analysis and evaluate them on the basis of what is plausible in your own evolutionary understanding. Come to an understanding that there are many obsessive entities and factors that want you to take the blame for what they are doing, and you will take the blame for what they are doing - in the astral; that is: warping and distorting their own psychic effigies into the psychic structures of others. They want you to take the blame, to get their own guilt alleviated and of course, you are going to take that blame. You are going to have yourself cannibalizing others; you are going to have yourself chopping people's heads off right and left and putting them up in pickling jars. You are going to have yourself doing all sorts of things as being the perpetrator, when actually, physically, you might not have done some of those things. And we are not saying that you haven't done all these things but we are just telling you to evaluate these things. Do not accept them on surface values; analyze them. 'Did I do these things? Was I that individual? Did I help do something similar to that? Am I making it into a distorted, warped monstrosity that is really residing in my psychic anatomy that has been warped out of guilt and frustration, venting this anger and tremendous negative force into my life by these obsessive entities? Or am I just wanting to be whipped and beaten, even to being the most negative of all negative people? If I was negative, I want to be the best or most negative! I want to have done more horrendous deeds than anybody else; that made me somehow better than the other negative people.'

"Now these are insane thoughts, yet if we truly analyze our own perspective and our own situation in these various relivings, there are a lot of things that come through our minds that are pure obsession, pure negative entities who want you to live in this subastral, distorted, demented condition so that you cannot really evaluate what you actually did do. For instance, the episode with Dalos: We might physically have just pushed the button, but mentally, what happened? All of these things were horrendously reproduced in effigy and imaging or imagery to create a real thing in the psychic. But to the person being tortured, it was very real! So in our own psychics, this was real and yet, physically, all we did was push a button!"

Michael: "It has come out that there were television shows and media productions of what was happening to Dalos concerning the physical contortion, trying to subvert the masses on various planets, using this."

Answer: "It was a made-up job; costumes, makeup, and all of the things that required an exact facsimile of a distorted configuration."

Michael: "Was there a significant level of karma incurred by the people who were watching this and cheering it on? You know, you will watch something on television and you are saying, 'Yes, yes! Bad guys!' So in this situation, I imagine people probably watched and cheered. Is there a significant part of karma that was incurred by these actions? Mentally, the thing is there; what do you do?"

Answer: "Yes, you are right. There would be karma induced by anyone who would voluntarily or involuntarily join the negative forces in negating the forces of positive energies. However, the force would be much greater with the person who was actually perpetrating these things, rather than the person who was just sitting and watching these things and letting them pass by him, and as you say, 'rooting it on'. The karma is instilled in both people. As a matter of fact, it depends on the individual involved; also, we cannot think in general statements about this and yet if you understand energy, if you actually maliciously attempted to physically accomplish this thing rather than sitting back and letting somebody else do it and to not be actually involved, there would be judgments within the self of how much was attempted to be accomplished. Putting it simply, it's a personal thing, as to how much is involved in his karma there.

"Now this thing about letting the people know publicly how negative Dalos was: We all know he was not negative; he was a very positive force. How did we establish the fact to the public to visually show how negative he was? Very easily. You could create all types of special effects and make somebody look just like that and actually have false, distorted configurations as in the monster movies; you know there is a nice looking guy under there but he looks like a monster. If you weren't told there was a guy under there, you might think, 'Gee, what a horrible, ugly, distorted thing!"

Stephen: "Because we had the ability to control mental processes, wouldn't we have been able to have Dalos do just about anything and then been able to film it? To discredit him, we could have him doing lewd acts. Would that have been possible?"

Answer: "I don't think there was any physical manipulation. I think that was all done with illusion. It was all done with actors because we didn't need him. He might not have been able to be manipulated like we wanted him to be, so we didn't take a chance. We just got an actor and told him what we wanted him to do and he did it gladly."

Keymas: "So you might say the capture of Dalos served two major purposes: to keep him out of contact with people and to obtain the necessary data that we could incorporate in special effects, reproducing his images as with actors. If we weren't using him in that way, we just basically wanted to keep him under cover, keep him suppressed. But we still did film him being tortured; we still did project the high frequencies and horrible sounds and beam pain frequencies toward that Being."

Answer: "Indeed so. Now you know and you all understand what took place. Nothing physically happened to Dalos; it was all a manipulation of static energy from the computer. Do you relate to that Steve?"

Steve Anderson: "Very, very much."

Calvin: "There was a concept brought up that we captured the other polarities too, and brought them."

Answer: "Well, that didn't take place."

Calvin: "That's what I wanted to know. Did we portray these other planets that did not join the Empire as being weird and not normal and all sorts of things, to the other people of the Empire?"

Answer: "We projected everybody being weird, everybody being inferior and then, 'We will overcome your inferiority and achieve making you one of us! We will help you become part of the Empire.' Does anyone have any more thoughts about that?"

Bill: "There was one movie out with a title like, *Jason Argonauts*, where he went through a mental crisis and felt like the ground was coming up underneath him, and that he was again falling. Then all of a sudden the ground would reappear underneath him again. Was that a particular movie from some past?"

Gordon Jenkins: "Yes, the Witch!"

Answer: "Well even with the technology that we have today, these things are very plausible and possible. They have truth drugs right now and all kinds of serums and mind-altering drugs and rays.

It is all very much within the science of today, yet the Unholy Six was thousands of years ahead of us in technical advancement. We are just little babies as far as the technical crews of the Unholy Six Planets were concerned; therefore we can well imagine the instrumentation that could make a person appear to be something that he is not, and then alter that state of consciousness to bring him out of that and show him that we really do have power over him; we can create anything we want around him and make him believe it and then bring him back and say, 'See, we can release you from all this anytime you want; anytime you will acquiesce to us.' There is a show called *The Sixth Sense* that shows this concept."

Keymas: "Did any of the positive, Lighted Forces or Brothers who had incarnated down, incur karma even though it was for a positive deed when they went into battle?"

Answer: "You people don't realize what karma is. Living the physical life, breathing every breath is incurring karma. What is karma? Karma is living the physical life and all of the things that the physical life entails. That is, even living in the most gorgeous, lovely, ivory palace, having all kinds of servants, living a philosophical life; living the beatitudes or the Ten Commandments or however a physical life can be, you are still incurring karma because you are living in the environment of the 3rd Dimension and living under the principles and laws that govern the 3rd Dimension; that is incurring karma."

Michael: "How do you work it out so you don't have to reincarnate back here?"

Answer: "By slowly reevaluating your idea of what the physical life is and what kind of life you want to live. That is the reason the man, Jesus said, 'I am in the world, but not of it.' It is because of your consciousness. This is what we are saying: 'Living the physical life, you have to be conscious of it. When you are shooting off a cannon, you can't be saying that you love those crystal cities on the higher worlds. You are thinking, 'I am going to destroy somebody!' Even under the circumstance of thinking, 'I'm doing it for the good of the galaxy; I'm doing it for the good of God and my brother,' you are doing it because your consciousness is in the 3rd Dimension, living that particular expression and being influenced by it. Only when your consciousness generates a desire to live in a higher, 4th Dimensional world, another dimension totally, from this one, shall you do so and this is why we have to study and try to conceive these energy principles.

"Those experiences we are trying to conceive to work out this past karma here, are rebuilding our psychic anatomy, our psychic body which really is constituted of energy of consciousness where our consciousness is. All those wiggly wave forms and vortexes are a force of consciousness and make it real to us because we have conceived the solidity of its value in our life.

"This explains why we call this the world of Maya and illusion, yet I am walking around and I feel solid because my consciousness, in my psychic anatomy, has solidified this arena or environment momentarily until I can rearrange the energy vortices in my psychic anatomy to be of a consciousness that is of another world."

Margie: "How were the people on this planet involved in that time?"

Answer: "That's a good question! Many of those peoples are living on Earth today and having karma in the stellar systems of Orion. They are now generating lifetimes in the Orion Nebulae and those energies could very easily blaze up again and we could attempt to reestablish that empire in a physical way again. This is why we have been more or less marooned on the outskirts of that physical environment so we can work these things out without actually reinstating the trauma of our psychic anatomies and empowering these astral obsessions that have been given a force into this physical dimension to repolarize the negative experiences.

"That is why we are on Earth today. We have more or less isolated ourselves with the help of the Brothers to work these things out. They give us these healing radiations so we can understand our past and unentomb ourselves. We can rematerialize the circumstances under which we incurred them originally in the dramas and ways of life of the Earth at this time in repeating the various histories of the Earth world on a much smaller scale as we are being given the opportunity to advance ourselves without decimating the entire galaxy! Is that understandable?

"This is why we are here: We have chosen to be here through our misdeeds and malevolence toward the Infinite as we have requested to be healed of these things and begged to be given the opportunity. The Brothers decided that this is a good time and place to set in motion the start of the changeover. There is a specific frequency you need at this time, in this junction and our karma is passing through these minor and major lines of cyclic force. So, 'Yes, we will give you this opportunity,' and thus, they have and are doing so!"

Jeff: "It is interesting to note that almost all the science-fiction assumes that, in the future, the Earth will be the center of some kind of empire. It is just our past. I don't think I have ever read a science fiction book that didn't take that for granted. The old religious religion books not only said that the Earth was the center of the solar system but of the entire galaxy and the universe. We felt it was."

Answer: "Yes, all these science fiction writers are remembering when all of these planets were invaded and this is why we see most science fiction writers relaying to their readers the idea that outer space, alien people are negative and destructive; that they are degenerate, superior-acting entities - all of the things that we, ourselves, were at that time. He is remembering his own past when he was this way and he wants to repolarize it and direct it into this dimension! There are some things that aren't factual; they are fancy, but yet not too far from what happens on the astral world."

Bill: "Edna and I have been questioning people and writing down what they think of alien beings, but we are now switching to asking, 'What do you feel about Higher Intelligent Beings landing on Earth?' And we take down what they have to say and this one artist said the exact same thing you said. You could see the electronic tuneback when we brainwashed people to think there were creatures out there, instead of High Intelligence. It is very valid."

Answer: "Yes, it is valid."

Helen: "As Jeff said, there is the assumption that we would eventually become the center of a big organization but then there is the other assumption that we are going to blow ourselves up. This is the other truth that we haven't pinned on yet. We did totally destroy ourselves when coming to the point of confrontation with the Light."

Answer: "Many did destroy themselves rather than be included in the ego-deflating condition of succumbing to another empire; they thought it was just another empire; they thought they were simply being defeated by a superior force and, 'We shall rise again!' So they very easily killed themselves off. Many thousands killed themselves off rather than be rehabilitated."

Helen: "I remember being told that they destroyed the mechanical planet that controlled the Empire rather than submit to the Light."

Answer: "Yes, that is true; many thousands of them did."

Helen: "Not as individuals, but as whole planets?"

Answer: "No, the forces commanding the warships, when a battle was lost out in space said, 'This territory of space, along with these planets, are lost to us as controlling influences; we shall leave behind those planets, whether devastated or blown up. We are not going to leave them to the Light; we are not going to leave the forces."

Calvin: "I guess this cycle started at night. When I lay down to sleep, I had all these images come into my mind and they are just grotesque! I was wondering if my concept is right. I say to myself, 'No, this is an obsession, I know that it is. Get the heck out!' Is it right to say, 'Get out, I'm not going to put up with this obsession from my past,' is that not a recognition and a realization that the Brothers are helping me with this problem or am I denying something I am not wanting to look at?"

Answer: "First you have to realize that thing is coming to you because you have tuned in to it somehow. You have tuned in because you have somehow negatively oriented your consciousness to bring it out. Now it is right there, and the mandatory concept here is that you have to look at it and see it for what it really is before you can say, 'I deny your existence.' First you have to realize, 'That's part of me - a horrible old icky entity! Get out! you have had power over me but now you no longer have force over me.' This is really exorcism. It is recognizing the force of it and denying it any more force from that moment on or from its exerting any kind of influence that would be detrimental to your evolution. It is not only words one may say that help, but the consciousness, as always!"

Steve Anderson: "I relate to this very much as far as your computer is concerned. I feel that I am seeing from experience but it seems like pushing buttons was very crude. It sounds like there is a button for snakes and there is a button for spiders. What comes to my consciousness is that this computer is a device that reads the consciousness of one person who sits there in a chair with some sort of device that is attached to him and it reads his consciousness as far as he conceived of something horrible - a horrible, lewd act or thoughts, or prophecy that is drawn from his mind.

"Then the computer computes these actions,

amplifies them and then creates them into a form of reality, projecting them into this force field to Dalos. This is something a little more sophisticated and can be directed by the person."

Answer: "This could have been in experimental phases but I don't think we were advanced that far under this particular circumstance. I think they were still machine-oriented and not projecting their minds because what you are talking about is really sophisticated. The machine has to take all the mental thoughts and correlate them and tolerate or understand what you are thinking to reproduce the scene."

Steve Anderson: "Well, if it could reproduce sufficiently to cause that effect in someone's consciousness in the first place . . ."

Answer: "All you would need are interfering electrodes in certain parts of the brain. All you have to do is interfere with certain activities because that is what ears are anyway - energies coming from the psychic. You are presented with some image or illusion and are automatically attuned to certain cells. Brains are alerted and they fire under that certain circumstance. They are alerted because of the vortex in the psychic. So all you have to do is play on the normal fears and phobias of most people and learn what they are. In fact, there was a motion picture on this and they put a probe in his brain and tuned in to various reactions and emotions, and he also had all those reactions. They could turn him on and off and eventually he went out of control and they could not turn him off and on and he had these fits of violent temper and would kill because he was tuning in every time somebody said something wrong. That would tune him in and he would go through that act and they couldn't turn him off.

"So with this machine, you could dial any frequency you wanted to go to the brain, and block off or create inhibitive factors. The way I feel about it Steve, was that you were using that experimentally and it wasn't under everybody's control."

Steve: "Well, that could be true; I was going to give a testimonial on that."

Answer: "Oh, you were working on that but the rest of us were working with these little dials; we didn't have the mental. You see, it takes a very image-minded person to project that. You just can't think about it, you have to create and regenerate within another's mind a directive force."

Steve: "A directive reality, yes."

Answer: "And then the one directing the force has to be clairvoyant and he has to realize these energy principles that are far beyond the abilities of the average person who turns dials. All one sees is a reading on a dial and anybody can do that - read a dial but putting this image in your mind and generating it and then projecting it into some machine which then again turns the image into a reality for another person - that's above the average mind, even on Orion!"

Mike: "So you couldn't have tuned him in to any more than he had within himself, right?"

Answer: "No, you could not, but everyone has gathered subconsciously through these many lifetimes, these fears of snakes and spiders, etc., because of the way of life on Earth. They may get bit by a snake once every several lifetimes."

Jeff: "So, in essence, is this machine a hypnosis machine? Is that the idea? The person is under hypnosis and sees what the machine tells him to?"

Answer: "That's right, it is a very sophisticated form of autosuggestion - not only to perform acts but to see things and make them real in your life. Of course, hypnosis has that ability too under certain circumstances."

Christine: "I still don't understand how it was hooked up to the mind."

Answer: "Well, it wasn't, he was experimenting with it."

Christine: "I can't get into my mind . . ."

Answer: "Well, first you have to know how the brain works and how it works with the psychic anatomy and of the effect of any interfering frequency generated. It just has to be generated to perform these inhibitive factors and conductive factors in the brain cells themselves. The brain cells have certain specific functions, and when they release electrical current, there are voluntary and involuntary reflex actions. There are also other cells in the brain that convey information from exterior stimuli in harmony with various experiences pertaining to some fear of heights or snakes or things like that. In harmony with that, it would generate a third harmonic which would go down into the brain and actually burst and which would be imagined in this fear or this phobia. This would be very real to him."

Christine: "Were electrodes actually placed on his head?"

Answer: "I think this was a force field that was generated and there was no need to have any physical apparatus here. This room was the inside of the machine, you could say, and the rays were coming from all points of the room and would focus on parts of the brain. That is very logical."

Michael: "I would like to agree with you but I . . ."

Answer: "Feel free!"

Michael: "You said something about frequencies that could be interjected that would cause spurious frequencies and wave patterns that generate harmonics which end up with the picture in the person's brain. If we had the technology to know frequency bands could cause these reactions and that those frequencies could be generated, then it seems that those frequencies could also be able to be received. Those would be the same frequencies, that if you were thinking about a snake for example, that thought has a certain resonant frequency. It would be something like an EEG type of thing. It would be very easy to have somebody extremely sensitive, one who would be able to pick up such high frequencies of thought patterns, so it wouldn't be like you were projecting this picture onto this screen."

Answer: "I didn't say that at all. Steve Anderson said that and he was experimenting on that type of machine. We actually had dials and all. We would turn a knob for snakes and turn a knob for spiders and turn a knob for throwing beer cans. We would turn a knob for different frequencies; if there was a reaction, we would know that we had a phobia there. If there was no reaction, there was nothing in the psychic and we would go to the next one and the next one and so on. There was no physical energy necessary because those energies were in resonance with the psychic anatomy, like you said, and the brain reinterpreted it as imaging and there would be a physical reaction - a fear or a phobia. It's like when an alcoholic goes on the wagon and he sees pink elephants running around."

Michael: "No, I just can't see that; it doesn't sound logical. You must have had a five by five foot board with a knob every inch to get so involved. There are a lot of frequencies involved there.

Answer: "You underestimate the technology."

Michael: "No, not at all."

Answer: "Okay, next!"

Brian: "How long a time period did this last? I can't accept that this went on for a hundred years. It seems like the logical thing to be done, but as you said, one of their reasons for doing this was to break Dalos down to join them and have that Force. But how long would this go on? Why didn't they just kill Dalos after a certain number of years?"

Answer: "Psychologically, you can suppress a person's subconscious and keep him alive and then not do anything to him for awhile. Another form of torture is fear of being tortured. This went on and on and on, pressuring him to do something, then removing the pressure. It's the old spy movie technique. It's very effective and the reason it went on as long as it did, had to do with the periods when nothing at all was done, just so he would wonder what was going on at the time - what was going to happen next. That is fear, and they were trying to play on the nerves - the waiting game; then the torturing would begin again in another form to try to break him down.

"You see, a person's will, under the torturing code technique is stronger, the longer it takes to break him down. They were being patient; they were doing the things necessary to break down the will of the person; to cause him to succumb, but he did not."

Brian: "It's like they didn't want to give up because of their egos."

Answer: "Right!"

Patricia: "If we or someone knew about reincarnation, we wouldn't want Dalos to die because he would reincarnate and work again with the Forces of Light!"

Answer: "That is a good point."

Steve Sandlin: "I'm still trying to conceive this thing that is mind when you spoke of a dial we turned with various frequencies. Now we just had this dial and were trying to emit a frequency. We would watch the reactions of Dalos and when we got a reaction, we knew there was something in his psychic. We didn't necessarily know that he was seeing snakes but we did know there was reaction there. There wasn't a dial for snakes or whatever, but there were certain frequencies that caused reaction in Dalos."

Answer: "We will get down to the nitty-gritties about how the machine works, but that is not the point I am trying to get at. But yes, I'm sure with the instrumentation, that which we consider and judge to be possible and what isn't is only because our conscious mind says that computer can't get anymore sophisticated than we see it out there. That is fallacy and not fact as it means that we put limitations on what is possible with our present science. This is what the scientists do right now; they say, 'This isn't so, so it can't be. It isn't so!' Then in the next twenty years, there it is! And then they say, 'I knew it was going to happen all the time!' I want to get to that point.

"If you want to make all of this sophisticated

engineering, it only means you were involved in building that sophisticated instrument. You want to know exactly how it worked and all the little relays and everything. Well, that means you were one of those helping to build that machine. Do you relate to that?"

Steve Sandlin: "Yes. The part I wanted to say is that . . . "

Answer: "You want everybody to know how great you built that machine; that is what you want them to know! And you want confirmation of how great you built that machine. That's okay."

Steve Sandlin:"No, it's not."

Answer: "It is for that reason. We are trying to help you overcome these things. If you are going to deny the problems, that's what you did then, and you are not going to get anywhere in your evolution."

Jeff: "Is it true that the reason Uriel, as Dalos, never capitulated is because all these concepts of snakes attacking, etc., had already been polarized in his Superconsciousness and the negative effect was nil?"

Answer: "When a person is oscillating on the physical side of life, he has to sustain these lower energies of the physical life. When a person builds up a psychic anatomy that has subconscious energies, then he has to be subjected to or submit to these energies to live in the environment. So this includes all of the survival instincts, sex drives, fears and phobias; the insecurities of the Earth life force are natural with anybody.

"So any person who is a highly evolved person and who dons the physical anatomy is susceptible and vulnerable to the manipulations of the people around him in the physical. But the psychic, spiritual self is not affected at all.

"Now you can take the body and tear it in two but that won't affect the spiritual body. It does affect the subconscious with pain, etc."

Bill: "I picture Dalos communicating, 'I'm okay. I'm captive here,' this is what I have conjured up." (Uriel: I would not have to tell them; they would know! They are a part of me and/or Dalos.)

Answer: "There was a communication as there is a communication between Uriel and the higher Brothers. There were attempts to rescue Dalos at various times. But this was the stronghold of the Empire. They didn't want to risk an all-out war at this time. This is what I want to get into because we want to understand that which happened between the time of Dalos and the end of the Empire.

"We will do that by saying that the Empire from the beginning, was out and out evil - just totally evil! The people of Orion and the leaders who came out of the system, developed themselves out of wrong decisions made for the progress of the people who were involved. Isn't it always the case? We have evolved to a certain point until we start making wrong decisions; maybe making decisions that are selfish instead of for the good of the Infinite as a whole and making them, perhaps inadvertently. In doing that, we might even rationalize that it is for everybody else: 'I'm doing this for your good!' But when it comes down to the point, there is a pinching off of that soul from his progressive evolution to one of total selfishness, creating tyrants, creating self-centered, selfish, insanely jealous people. They are jealous of their position, of their power.

"That is what occurred. Masses of people were really diverted by wrong decisions made by themselves, and allowing the few to rule them under the auspices of being helped as a total configuration, the total civilized community. Then the leaders, making more wrong decisions, then began to use all of the energies for all of their commitments for themselves, instead of for the whole.

"Now when these energies are turned around, it is very easily seen that the person who is positively progressing and then reverts, still has that intelligence with which he arrived positive, but now that same force is being used introversively, or for 'self'.

It is still the Infinite; it's just how it is now being used.

"So it is being used for self, therefore, the intelligence is still there but now it's for self, so losing its infinite contact. In using that intelligence for self, great cunning comes into these things, great abilities to orate, to subject people into insecurities and manipulate them with these fears to the type of life you want them to live under you. So these leaders, making more wrong decisions for themselves in their evolutions, generated the Empire and all of the people went along with this; all those who were involved with the Empire.

"Now there were certain decisions made by the people and the leaders. (The leaders weren't the only ones. The people cohabited with them, endorsed them, venerated them as being 'holier than thou' leaders and set them up on pedestals!) That only helped the leaders to become more ego-seeking, self-god-seeking! So there were decisions made: We are superior, we have proven over this period of time that we are superior. Why not use our superiority and go out and help others live under us? This was the rationale. So the great movement was begun by the star systems, and it spread throughout the star systems very, very slowly at first, over hundreds and thousands of years and each system took its place in this confederation.

"So it came to pass that long before the Higher Forces Confederation touched them, physically speaking, the Unholy Six Forces began moving out of their territory into other star systems, offering their way of life which was very highly advanced, to people under whatever circumstances. They were not concerned as *Star Trek* was, 'Don't alter the state or the conditions on the planet,' and that was the prime directive. 'Let the civilization grow!' These Empire people weren't concerned with that; they were going to civilize that planet and they were going to force all the ways of life on that planet because 'they knew best'. They had proven their survival of the fittest and they had proven they could exist under the conditions they were living under and felt it was high time that the rest of the galaxy knew how to do this!

"So they went to these various planets and were actually very friendly at first. They said to these planet leaders, 'We are very advanced people; we are highly technical and we are very wise. We would like to help you become our subjects because you need us. You couldn't exist another ten years without us! We'll show you our great power - zap-zap-zap!' Of course they are in awe while we are saying, 'You see, we told you!' It's the old protection racket game. 'We will protect you from all other invading forces! We will give you everything that we have, if you will follow us and become members of our society. And whatever you have to give us, then you will give us tribute to supply it between the other forces of the Empire.'

"Some accepted under duress; others accepted willingly. It was determined by their state of evolution at the time, and wrong decisions made by them to join. So eventually, these long tentacles started extending out and there was even a point in time when the 'Confederations of Light' touched the Orion Empire.

"Of the peripheral planets that were touching each other, it was like two circles flowing together. There were a series of planets; one planet we are now trying to film for the program is Idonus with Macinus as leader. This was the period of time when the two confederations - the material and the spiritual just met - when One, Dalos came into the Orion Confederation and started planet hopping into the central portion of the Empire! Now the central portion of the Empire got wind of this - that there was an envoy of a new confederation - a very powerful confederation which was infringing on our territory. So we had to do something about it. They were like enemies to us because they were too powerful for us to coexist. It's like the *Clingons* and the Confederation of *Star Trek*. I like to use this analogy because it gives us a visualization of where we are.

"And so the central hub of the Empire decided to get rid of Dalos, as an example even, to persuade Dalos to come over on the Empire's side. Because by proving to everybody else that we were superior, we could prove to Dalos that we were superior. But we had to go about this in a sneaky manner because that was a very highly technical civilization from which this Being came. I do not believe that we consciously knew the high, spiritual status of this Being but we did know that with spaceships there was a very highly technical crew and entourage that was coming. So they were the enemy - the out-and-out enemy, we felt. So we had to take steps, which we took on the next planet on which Dalos was going to arrive.

"In the filming we went through the whole thing that we are reliving now and this is what happened: The planet leader, Macinus (Dennis), went to one of the outer planets that was touching the frontier planets of the two empires, you might say, and faked being Dalos (a male) because this planet was being touched by this lighted confederation, and this whole lighted confederation knew of Dalos. So if Dalos were here on the frontier, that would keep the Forces of Light at bay and there would be no alarm sounded within the Forces of Light camp to come to the rescue of Dalos. This was the logic. I don't know if it was the greatest logic but this was their logic.

"After this happened, the person faking or imitating Dalos, stayed for approximately four to five-hundred years on the planet Idonus. For one to extend the life stay was no problem.

"Idonus was a way-station for the Empire's spaceships and when they went past that, they went into no man's land to conquer new planets. This way-station was a pleasure planet where the crew members stopped off and had their last fling before they would go out past the Empire's realms.

"Now the Forces of Light found out about this after many years. They had a council about a hundred years after Dalos was under mental sedation, and there was this war that began between the Forces of the Lighted Confederations and the Unholy Six. At this time, the Confederation had about a thousand planets compared to the Empire's possessions, of about a hundred planets. So the Confederation outnumbered the Empire immediately and that is the only reason they eventually won; the forces that were pitted together, technologically had the same forces. Anytime the battle ensued, there was an even battle other than the superiority of numbers of the Light. This is why the Empire was finally encircled and whittled down, just like in Germany. However, just like Germany, the Empire said, 'We will burn our bridges behind us. We will leave every planet that we lose to the Light in ruins, devastated, distorted or maimed and blown up! If we can't do any of those things, then we will think of something else!'

"Whole populations were blown up and destroyed; the entire populations were smitten down during this period that lasted over four hundred years. At this time, many people were killed and even reincarnated again on other planets and again took up the weaponry under other circumstances. Finally the Empire was decimated and completely eliminated as a space force. This was the only reason: It was because the Forces of Light Confederation had much more psychic or spiritual force. The tremendous, negative, swirling vortex within the Unholy Six as an Empire was a tremendous negative force in this galaxy . . . just tremendous! It took a tremendous, overwhelming Lighted Force to overcome it!

"So if this total, overall chronology helps, maybe we can gain a little understanding now of our position in those lifetimes and what really took place here."

Mike: "I am trying to relate this to the last

Unholy Six Cycle that we went through with the Immortals. Is this the same cycle?"

Answer: "This is the same period, yes."

Mike: "So they also had the force field in this time period?"

Answer: "There were only a few of those who had learned the force field secrets. This took place long before Dalos came, and they actually helped themselves to the leadership roles.

"So when they withdrew to the Inner Circle, or wherever they were now putting their stronghold, they destroyed what they had been influencing for these many hundreds of years. If they didn't have enough time to destroy the planet, they would decimate the surface or they would distort the people there, whatever the conditions were. We can say that there were hundreds of different circumstances; different situations."

Keymas: "Is this what happened in Idonus the very last few years when the people were decimated?"

Answer: "Yes, and that was one of the first planets that went to the side of the Light."

Keymas: "In *Star Wars* I always felt that some of the beings they conjured up in that saloon scene were very grotesque, bizarre creatures and were just remembrances of the mutations. We, as human beings, are basically in a similar form throughout the galaxy. Is this true? Does one go into this two-headed, four-armed thing because something has been tampered with? Is this true?"

Answer: "Nothing is impossible and those forms could exist in a physical way but the physical form adapted environmentally on physical planets, would be in the majority. There might be alterations but there would be a trunk and limbs and head for metabolism and ambulation of the body; thinking processes, etc. There are probably numerous deviations from the normal stature, but I think the majority of people had this type of frame or structure. There are all kinds of forms though that can be considered man because of the evolutionary force they are going through. Are there any more questions?"

Stanley: "Would it be true that Communism in this country is a regeneration of that time because it goes along the same lines?"

Answer: "Every war is a regeneration of that time; every subversive condition is a minor or major regeneration of that time, you might say, because there is nothing new under the sun, and this is true speaking even of the Central Sun! Even those in the Unholy Six weren't making anything up that was new. It had gone on for as long as the Infinite had gone on in various sections of the Universe - some even in other galaxies spread even further than the Unholy Six before they were checked. I would say there would even be entire galaxies that succumbed to the evil force before it was neutralized (so-called)."

Michael: "So it has gone from galaxies to solar systems to squeeze these negative people all on one planet?"

Answer: "No, no, no! I am not saying that all of us are the drippings of an entire evolution of evil condors. What I am saying is that the evolution of man has the potential to be negative for certain spans of evolution and we can go one way or the other. But the Infinite is infinite; it is one unit force but it is how it is used at any particular moment that is so important. And a person who lives in the 3rd Dimension, always has that potential danger within himself of reverting to the lower elements of his nature; that is, survival of the fittest and that the ego is supreme and perfect over all other forms of the Infinite. That is the danger, the potential danger that is always with us while we are in the 3rd Dimensions.

"Many succumb to it many times before they finally crawl and creep and scratch and elevate themselves out of that potential danger to evolve into other states of awareness and states of consciousness that are equally as challenging, but in a different way.

"So this is the process of evolution. You are never secure at any one moment, of being exactly what you were in the last moment. You are always changing. In other words you are not going to sit back on your laurels and say, 'I have been this positive for this long, so now I can be negative and roller-coaster for the next few million years.' That is not the way it is done; you have to keep your eye on where you are situated at any one particular moment. The further a person progresses, the more stringent the rules become, you might say, or the greater the room for error becomes. It is a straight and narrow line that you begin to walk and as that road becomes narrower and narrower and then even more narrow in consciousness, there becomes the equally potential danger of deviating from it because you still have the psychic-widened perspective influencing you. You want to narrow your energies into a selective mandate in your evolution and yet you still have the dominance in your own consciousness of the ways of the physical world; the ways of the physical life, the survival of the fittest, the drives of sex, the drives of hunger and defense; all the mechanisms that sustain you in this road of life - the evolutionary course in the material world.

"Now you must narrow all of the thirst of this life down into a thirst for another life. While you are doing that, the danger is the greatest because the danger is within yourself in deviating from that narrowing course you are trying to pursue. Eventually, that road becomes nil and nonexistent because you are no longer walking the material road. You have attained the 4th Dimension and therefore, the wide or the narrow does not have the influence on you that it once did.

"So all men, whatever galaxy they are in, whatever universe man is in, has that potential to become infinitely, creatively minded. The choice is always his. When he makes the wrong choice in a particular point in evolution, his progress is impeded; it is momentarily reversed until he can compensate deed-wise from that judgment he has placed upon himself as being negative. When he has reversed that, he again becomes progressively-minded and oriented and for the moment at least, the potential danger has been avoided. But it is always there from one moment to the next and we must learn to cope with it.

"In this past, in the Orion Civilization, it was not coped with by the numerous multitudes who were involved with this energy state of existence. They therefore collectively, en masse, degenerated the life style to a type and way of cohabiting with one another on a very primitive level with very advanced technological instrumentation. When you combine those two together, you have a chemical synthesis that is very explosive and can predominate the human society to such a position and such an extent that it can very easily self-destruct, because the whirlpool of negative energies is sucking down into the center of self, all of the positive energies in a reversed fashion and each soul is imploding within himself that progressive impetus and negating the purpose of his evolution. Collectively, this was established in Orion; collectively, it could be established at this moment in another galaxy with a different set of souls.

"There are other hierarchies, there are other Fountainheads. The Infinite is just that: boundless, limitless, endless, expansive, total and has the potential to be all forms of itself!"

Keymas: "So it is very conceivable, if we continue to progress and eventually become a master or adept, that the group of souls who are incarnating on this planet who are comparatively young souls, could again, through the wrong evolutionary course, recreate the space-war routine as being a part of evolution."

Answer: "As we said, anything is possible, but we have to remember here that the souls penetrating these negative acts who reverse that bias of degeneration and work out their karma, are aiding not only themselves but all others; in other words, neutralizing that force within themselves. They act as stabilizers with these energies that are permeating the universe. When a society or group of people tune in to those, they are not only getting the negative aspects of those dastardly deeds done, but they are also getting pulses from the souls who are advancing and are now advanced and are warning them not to do it.

"So they are getting both sides of the coin, and it isn't that unbalanced now, whereas with we who still have not polarized in the most positive aspects, this Unholy Six reliving and all the others, there is a potential danger of many millions of souls still tuning in and activating in a physical form, these negations in any part of the galaxy. So this is very serious for our evolution and those who come after us.

"If you propagate a deed, an act or thought, you are responsible for that. You are not responsible for another soul who tunes in to that, but you are responsible to the Infinite in your own guilt agitations developed by the superconscious self, to take the responsibility for others that you have laid upon them the potential danger by ignorance within your own soul in not knowing how to be progressive. You laid this out into the Infinite as an active knife for any baby to pick up and stab himself or any other being. You constructed the concept and the idea of that knife as a lethal weapon. If you do not, within your own soul, reverse the bias and neutralize that weaponry concept of form you have generated during that particular part of your evolution, then it will forever be a potential danger for other people. To be progressively minded, you should and you must polarize the opposite side of any negative act you incorporate or imbue into your psychic body.

"So to polarize that knife as a lethal weapon, you must also polarize that knife as a constructive instrument. So when a little child should come in contact with this energy and begins to play with it, he will also play with the positive energies of it. He might cut himself a little, and he might even explore cutting another a little bit but eventually the overriding positive bias that you maintain with this thinking pulse from your own psychic structures to that energy that you have generated will help him. He will learn this is the true purpose of it, the positive side, and therefore if positively minded himself (and most souls are until they totally cut themselves off) he will see the necessity and indeed even have the desire to use it progressively and positively.

"So all of these things, we should think upon as we are deep in the middle of these relivings and with all the reactions and emotions, continually see the opposite side of what we are going through. It is the character builder and it is the potential energy formation in your psychic body replacing the more negative forms so that you may live and exist and cohabit with beings who have already arrived there, shedding their overcoats, their emotional vicissitudes.

"The kernel of their soul, the lighted flame that you now see them as is their true self, their loving self, their real self, their Infinite self. They will see you also thusly and treat you with respect and with loving kindness - not as the souls you encounter today in the environment you have - the reactionary emotions, the problems, insecurities, etc. The jealousy and resentments you have for each other are born out of the slime of the egotistical mud, the ooze of the beginning birthplaces of souls of the individual being. This is where the ego originally developed and this is where it must remain. It cannot journey into the higher worlds because respect is gained from humility toward the Infinite - a serving of all energy beings and intelligences in that Infinite. This is what gains respect from other intelligences.

"If we wish to be respected and be a participating part of these higher dimensional aspects, these abstract dynamics of the Infinite Intelligence, totally aware of itself, then we must shed ourselves; we must levitate ourselves from this slime in the primeval ooze of the reactionary ego self. To do this, we have to reinstate these deeds that we ourselves deem negative, destructive and evil and face them for what they are: a mental construct in materialization into the 3rd dimension as a real platform or a pedestal or a vehicle for us to maintain our insecurity in this world until we could learn something better. This is what we have referred to as the monsters, the negative obsessions, the entities that claw at us in the night; the slime and the ooze; a formless blackness that surrounds us.

"All of these things we have needed, we have clung to, we have indeed become one with to survive. Now we no longer need that arena or this arena to survive because we have quested and are learning to face those negations, those things that sustained us for what they really are - the illusions of reality, a vehicle of consciousness. They no longer have power or influence over us; we do not allow them thusly to do this to us.

"As the Moderator says: 'We dismiss them, and they will fade just as the mists fade before the morning sun.' They must, for they were born in the nighttides of consciousness; they were born in the marshes of an Earth world existence and they were born out of ignorance of the life-sustaining principles. Thus, with the resurgence of new worlds, new knowledge, new wisdom imbued and impinged in our new psychic structures, so will we shed these encrustations and become cleansed! And in this cleansing, we will show our true self, this higher self that we have worked very diligently and long and hard to attain. In the attainment thereof, so will the respect come - not necessarily an ego-builder but indeed exactly the opposite; the respect for the Infinite and of the Infinite for itself; for has not indeed the Infinite justified the existence of itself? Even as the cocoon sheds itself into the beautiful, radiant, soaring butterfly! The Infinite has justified itself in our own evolution by proving the progressiveness of itself, the strengthening of its positive mandate. That small flame, that flickering light that now walks in the higher spiritual dimensions will grow into a lambent beauty - that butterfly, the iridescent and opalescent Flame of Life.

"And thus written and charted within that flame are all of the experiences that are needed for an even yet higher flight into the Infinite. Thus, dear ones, we leave you not, for you are our love and your journey has been ours; your sojourn ours. Look up! Look up to us and the Ones who have shed these shackles and have freed ourselves of that which you call 'material life'. (Tom weeps.) (Uriel: "He should;he has not released his past with this destructive Orion Cycle!")

Uriel: "And thus, so shall our Light and Radiating Energies come to you as you travel toward your star-studded future!"

Channel Uriel: "The foregoing information was channeled from the Brothers of Space and Light - Leaders of the other planets, thus they are in a good position to view all situations and circumstances on this as well as other worlds due to their Infinite Consciousness attunement with all worlds. These Beings are not only those living on physical worlds but those of worlds far beyond the need of atomic structures; worlds and people who no longer need or have physical bodies. Thus we can receive their words of wisdom in full knowing that they speak from experience for they, too, once traveled this lowly road and overcame the way of the Earth people.

"And so we could say, 'Heed ye well these words of wisdom which they have been so generous to share and let us make the most of it in our day by day lives. 'The proving of these concepts lies in the benefits gained when Principles are worked with and applied. The results are as certain as night following day, when concepts are thoroughly conceived."

Chapter 10

Testimonials

June 13, 1979 Meeting:

Dorothy: "I was so transcended with the dissertation from the Brothers that yesterday has sort of lost its reality and there was a wonderful healing. So I will tell it in an entirely different way, I am sure, than when I came in tonight. The Dalos Cycle has been extremely strong for me because I have been with Uriel and I have been unusually irritable and irritating. I am sure she would bear this out because I have been using those horrible energies from that cycle.

"It started with little things like spilling clorox on the rug and I tried all sorts of things to get the odor out. Everytime she went by that spot, Uriel would get that horrible chemical smell so she finally said, 'Dorothy, if you don't recognize it, we will never be able to get rid of that odor.' Of course, it was the nasty smells projected to her during the Dalos cycle.

"Then there was this heat spell that was unbearable - at least it was unbearable to me. Uriel took it in stride and didn't mind it but I was all upset with it and miserable. That, too, was an attunement to the rays of heat and the rays of cold directed to Dalos, so I did recognize that.

"There were so many things that just kept going on and on, until I was really almost ashamed to face Uriel. There was the food which became evident yesterday when we came to the Center and stopped at a restaurant to get something to eat. While we were there, she was talking about this list I had typed of the negations we students could portray in our reliving to do to her - to do to Dalos. On the way home I became so sick to my stomach I didn't know if I could make it home. I arrived there and lost my food in my nauseous state. That represented the poisoning and rotten food that we fed her, as Dalos, so when I tune in to anything poisonous, I am in double jeopardy because of the old Pharaoh Cycle. There is a lot of poison in my psychic to get rid of yet because I have it now in the body. As I say, when I came in here, I was really sick with the whole thing, but this wonderful feeling tonight with the Brothers so close . . . it was diminished and I'm sure I can gradually work it out from here."

Kathryn: "A couple of pieces just came together. I knew we had fed these venereal-diseased bodies to people and this is why I feel that I fed the diseased people to Dalos. I was caught in my mind and I couldn't understand why we were feeding people diseased bodies when we had good, healthy bodies; we had regular factories going for food. So this has cleared this up for me."

Christine: "Boy, this cycle has just been almost too much for me to take. Tonight I tuned in to really how horrible it was and I should say the mind reading the Brothers gave me a few seconds ago, because I was told there were other people involved here. What happened to me was that when Tom started talking about how the negative leaders, when they would leave a planet, would burn their bridges behind them in the manner of blowing up the planet, that tuned me in thoroughly and I was feeling like I was going to explode. I seemed to be getting more and more insane and when this hysteria started coming in, I couldn't control it.

"I then raised my consciousness to ask the Brother to please help me and in this act of asking for help, this 'Please save me,' type of thing, I tuned in to this one particular past lifetime where we were all involved.

"We were stationed on a planet and were negative at that time. We had been playing havoc with this planet and the Forces of Light were coming, and we had to get out fast and when we were boarding this spacecraft, which took off somehow, I was left behind. I knew of the plan to explode the planet after we left, but I was there. That is what I was waiting for, sitting in my chair over there. I tuned in to this horrible fear that I wasn't going to make it, and in that lifetime, I didn't make it and here I was, waiting for the explosion. Right now I felt that I was going to explode and there was no way to stop it! I have never been tuned in to anything in my whole life so thoroughly, and I know what it is to be tuned in now!

"Up to now, I think everything has been superficial and was nothing compared to this small thing that happened to me tonight which will prove that I have been shown what is in my own psychic. But being the victim is nothing compared to what I have done to other people - absolutely nothing. I get to the point sometimes where I don't think I am going to make it. (Chris breaks into tears.) But I know that's not right, if I would just think correctly; I have had many things shown me in the last couple of days that have given me an attitude, futile in nature, which I have right now.

"Dorothy told me something that is always good to remind myself of, 'We are never given more than we can handle.' Many times I think I can't handle it but I really must be able to if I am seeing these things. And about the pain we inflicted on Dalos, I realized I had a very big part in it tonight when I looked over at the dials that Jeannie is working with. I said to myself, 'That looks like the pain controls - the pain threshold.' Back in the time of Dalos, we would just get so far with a certain hallucinogenic pain and then we would hold it there; not too much and not too little but we would keep it right there. I have been reliving this while I am varityping because there is this dial that I can't go beyond or it will make the lines too long, but if I go under, it won't fill the pages, so that's not right either. I'm always manipulating this pain dial!

"Of course, now I am working with a psychiatrist

who deals with pain; all he deals with is blocking off people's pain that he created before in previous lives, but I created it, too. I almost walked out on the job today; it was just getting too painful for me but I can't do that until I work it out. So I am very grateful to the Brothers. It's a miracle that I am in this place now in my evolution where I can face these things."

Patricia: "Last Tuesday, the Valneza group went to the mountains filming and I contracted poison oak which caused little blisters on my legs. For the past couple of days I have also been having a bloody nose. Well, of course, there has to be some cause for this to be happening. It came into my mind that in tampering with the magnetic fields, we could use certain high frequency radiations and through hysteresis make the person bleed through the fingernails, the ears, etc. I was thinking about this being portrayed in a movie called *The Fury*. When I would get a bleeding nose, I would realize this and it would stop. But the one thing that was also in my mind was that we would do this to a particular portion of a planet.

"I was set up as a sort of 'prophet-type' person and would say, 'You are not producing like the rest of the people. God is going to punish you,' and then this tampering radiation would be activated. The thought also came into my mind that I was left on this planet, although I hadn't relived it. Then I came down to the Center and stepped out of bounds.

"When the songsters were practicing, I said something. In my mind, I was trying to be helpful but I was just butting my nose into it. I then went down to get a lemonade for some of the students and when I came back, everybody had left for class and I was here alone, carrying all those lemonades. I was resentful and hateful, 'It's their fault! It's all their fault for leaving me.' Then I went to the parking lot and saw Anne and she said, 'Well, you know it's all your fault,' and drove away. (She wasn't going to the class.) I was so emotional at that point that I thought, 'I'm not going to make it.'

I ran up to the Lab and had these terrible pains. I came to where the costume rack was and just sat there crying because I saw this emotionalism I had towards everybody.

"I wondered, 'Am I ever going to make it? Am I ever going to be a Light Bearer with all these emotions that I have?' As I sat there I realized that I was reliving trying to get some sort of protection from radiation in something like the bomb shelters that we had made but which were broken down and of no help. I had this writhing pain and realized that in the past, it wasn't so much that we would make them bleed but there would be so much pain and most of them would survive and would not deviate from their job again. I think I had such a psychic shock in that lifetime that I did die from this.

"But the important thing about this was that when I just wanted to go home and lie there crushed, I said, 'No way! You have had a realization; you are not going to lie down with it. Get back to the Center and start typing or doing something positive in the only way you are going to get out of this emotionalism you are constantly going through! And it helped because all this hatred I have had towards the students was released. I accepted that it was my fault and, 'Yes, they are reliving something too and learning,' so I stopped beating on myself. More times than not, I negate any realization by beating myself and saying I am worse than the dirt of the Earth.

"So just by realizing that we are reliving we can use this as a positive action. I am finding out now what progressing is all about."

Thelma: "Just having returned from the planet Valneza and finding out what a big facade this was, I find that my whole life has been just one big facade. I never really found out who I actually was. In the Planet El and also Planet Valneza, I am seeing in the background, dissension or throwing in little lines of distrust. I have never been able to completely come out with my true emotions and tell people how I really feel. I have always waited until someone else expressed an opinion and I would either agree with it or tone down my opinion, depending upon what I wanted out of this person. I always wanted to be liked and I never wanted to take any positions of great responsibility.

"I guess I realized that from my past my guilt is too great but that didn't stop me from trying to control things from an undercover standpoint. I have always tried to insert little things here and there that would undermine something that I didn't go along with. I find myself in many situations as sort of an undercover agent. I never would allow my true emotions to come out unless it was something that I really strongly believed in. In Planet Valneza, I worshiped the goddess of flowers and when it came time to destroy these, I chose chemicals. I realize now, that I have been involved in chemical warfare in this present lifetime, although in a supposedly positive way (I, being a nurse). I am sure that you have all heard of mustard gas that was used in World War I. That has been refined and is now given intravenously to cancer patients. I have given this many times.

"When this cycle started, the bottom leaves on all my plants at home would start to die; especially my big plants. I was starting to worry about it but after I realized this was due to chemicals I had used, they are no longer dying. Also, when we went to film the flashbacks in the lab, the makeup was a surprise to us. We really didn't know what kind of makeup we were going to have and that was a workout in itself. I finally realized why I wore this red wig. It was combed very straight to look severe. I realize now that all my life I have never worn my hair straight except when I was a child. I have always had curls around my head and I feel uncomfortable with my hair straight.

"Over the weekend I went to Disneyland with Decie and that was quite a workout too. Standing in line to get on one of the rides in the afternoon when it was hot, tuned me in to the people being led to the slaughter houses. It was very nauseating and then we got into the people-mover ride and into the sound tunnel and it broke down; it was like a force field really, and all this music and all these cars and movies were coming at us. The sound was just horrible. Finally, they did get around to turning the sound down and that tuned me in to the horrible noises in the force field."

Dennis: "I haven't known who I was because I have been impersonating other people and this is what this cycle is all about for me. I was the fake Dalos for about 500 years on this planet. This is a really devastating cycle in one sense but is really a great opportunity for me to finally move in the right direction now and leave all these facades behind. I have had this guilt with Uriel for the past few weeks because of course I have been confronted with the real Spiritual Being that I tried to portray for negative purposes.

"With the dissertation tonight, I saw on a more expanded scale how I was that cog in a negative wheel that was being perpetrated at that time and how taking on this facade allowed the Empire to build its mechanism, its juggernaut while this fake Dalos was out playing God. The insecurities I must have been going through in that lifetime have been coming through because I supposedly was to have been the one directing this play, but I have been such a jumble of confusion with no coherent thoughts that here I am again, the puppet leader in name only! So this is going to be a real tremendous cycle to work through and work out.

"This morning I relived one aspect of it and that is the electronic surgery I underwent to look like Dalos. I went to the dentist this morning and had a cavity filled which related this cycle to me. I need a bit of bridge work done and he referred me to another dentist to extract a tooth; the other dentist is an oral and maxillary facial reconstruction surgeon. My dentist handed me a card and I said, 'Yes, this guy starts the work and then I go to this other guy and he finishes it off!' I am totally inphase with this reliving and all the fakery.

"On this planet, we offered rejuvenation to the people of Idonus and lasting beauty to those who would follow the Emperor! So with this plastic puppet up here playing Dalos and saying, 'Come on, you can be an immortal, you can be beautiful, you can join the Empire. Just step right up into this machine!' Then there were bells ringing and lights flashing which were very attractive to this planet and helped bring this planet to the negative side.

"When the positive forces did come to this planet, some switches were made and the Orions pulled up stakes and left the fake Dalos behind, and they reversed the bias of these machines and many of the people became mutants. So in that lifetime (and we are going to portray this in the flashback sequence for Idonus), I was killed by the people who became mutants. I can't think of someone who deserved it more but it's a lot of karma for the people involved. I had an image flash into my consciousness tonight that showed when I first came to that planet and walked out into the villages - just playing it to the hilt, telling them how I saw their auras and how beautiful they looked. I played this sick person, a totally insane person trying to appear as a Brother. I saw the adoration of the people who thought I was for real and it left with me such a hollow, empty feeling for all these things I have done to Uriel in the past; this is what I was seeking, this adoration and admiration.

"Now I have seen it for what it is and how empty it is. I know that entering this cycle and having the opportunity to work out this particular past is really going to help me get going in the right direction. Thank you."

Bill: "I had a realization presented to me about a week ago, but the Unholy Six brought it in where it started, I felt really young; I have gotten into the military a lot of lifetimes when I was very young and this is where I have been very ambitious and don't really have good control of any logic and reason.

"When I was in the military in the Unholy Six, I was very young and in this one lifetime very ambitious to do whatever was going on. What was really going on was when we were looting and taking their women and goods, I was right in there and I loved it. I relived this before in this life and I recognized it in the time of Zan. I would go to other villages and bring people back whom we would take from other villages. This was the way I treated women in this lifetime; I would like them to a certain point but then if I were really riled, I would have a strong feeling in my mind that I was better and would treat them like a dog - with no respect. I know this came from taking these women for our pleasure and then to the pleasure planet. I would take what I had stolen and keep the best for myself if I could. The things I didn't care about I would use to bargain with. I would bargain to have a girl for a week or bargain to be able to get this particular fantasy on the pleasure planet. I feel like I had done that.

"When Dennis spoke, it rang in a note. When I take a snapshot of Uriel sometimes, I get a little carried away. Sometimes I get about twenty photographs of her and I leave out the group. What I really feel is when I handed her the pictures today, she said, 'Does this really look like me?' To me it was a beautiful picture; she really looked thirty and radiant. I said, 'Oh, it looks just like you. That's the way we see you - young and beautiful. When your higher Self comes through we see you as beautiful, as a thirty-year-young mature woman.' When Dennis spoke, I said, 'Oh, that's where that word fits in, 'Is that really me?' I tried to make Dennis look like Dalos when I was filming in the Unholy Six. I really feel strongly that this is where I picked up a lot of hatred and resentment because every now and then I become so irritated when I am around Dennis, for he probably was fussy about how he looked and how everything about him had to look - just perfect. So, being guilt-ridden about what we all were doing, I incurred a lot of karma with Dennis that got mixed up in other lifetimes too."

Christine: "I really feel better now and I want to relate this one realization I had of a past in this Dalos Cycle. I don't think I was one of those who was made up as a replica of Dalos, but I certainly was involved in a situation where there was a court, and a replica of Dalos would get up on the chair. I didn't play the part of a judge, but I think I played the part of a lawyer and I would strut back and forth making the fake Dalos confess to things that weren't true. So, during lifetimes after that, I have put myself in the same position of getting caught. By getting caught and doing negative things, I, myself, would have to go on the jury stand and not being able to defend myself because of my great guilt, I admitted to the things that I would do wrong.

"This, to me, has made me have a healing of: when somebody points up something to me, even though I may not have done it, I will admit my guilt because I am so used to admitting my guilt when I would be punished. It's almost like not being able to live up to your own convictions and so we feel sick about ourselves and feel weak all the time. So I am trying to figure this out thoroughly so I can reconstruct some kind of foundation in my own psychic and not be such a wishy-washy person."

Kathryn: "My insides are really pounding. Two days ago I saw Paul, a student who used to come in to the Center. He had just gotten out of a mental hospital. He shook my hand and I said that we were filming and he was welcome to watch, to come over. This attuned me to many obsessions that were very low, maniacal-type beings and also to my lower self when I was in a mental hospital and just a completely insane being, with many thought forms I have built that are a part of me.

"These things are surfacing within me and so the only thing I could do was to scream to get this out. I kept driving into the mountains, screaming. A few times I was just totally involved in the screaming and then I became aware of the Brothers and I could scream and it was like obsessions were leaving me through my throat and stomach, the different chakras in the physical. I knew these obsessions were connected to me by frequency relationship and were taken to the inner to the hospitals. So this is really great that I am a channel for the energies to be changed.

"This was very realistic to me and very sane, and my mind was able to comprehend and understand this. This went on for quite some time and as I was looking in my rear view mirror, I saw a burst of many different colors of energy and it just stayed there, pulsating. Then I moved my eyes up a little bit in the rear view mirror and a little dove was flapping its wings. I said, 'Well, I would just like to fly to be aware of the Brothers at all times and attune to them in this way and have this clear understanding of this part of my being that is expressing on this planet.'

"I felt this sanity and this awareness; the reality of a clear consciousness of what's happening. So this little dove sort of stayed up in the air and followed me. I was very transcended and then I saw this ball of Light shoot down and then another ball of Light. This kept up for a little while and then I became unaware of everything around me. I knew this was the Brothers. Then I heard some dogs barking and the lighted configuration disappeared. I became involved in this screaming the next night when a lot of things were coming in again. It helped me and was a good way of releasing a lot."

<u>Jennifer:</u> "I would like to thank Uriel and the Brothers for giving me this opportunity in the Valneza group where I had a chance to see who I really am and to see this facade that I have erected. This person goes back to the Unholy Six and doesn't know anything but love of self and conniving and is just an egomaniac. It's one thing to see glimpses and it is another thing to be in that

insane consciousness and it's going to help me.

"I have only begun to see these energies because I live with them all the time and I know that this is I. I am still in a cycle with the Dalos thing because I have great insecurities when I see Uriel and I have this great facade and I am very fearful. I have been doing a lot of typing but I become frustrated at it. Something always goes wrong when I do it; either the microphone or the tape and I know it is because of what I did to her - the pain and the insane pictures. Some of those pictures have been coming back to me and I can really relate to what Tom said earlier about leaving her fine for many years and then we would start in again. I will go through these states of consciousness where everything is fine and I will feel clear and can work; then all of a sudden I will be in this insanity and I will have to start reading; then in the dream state I will see some of these things.

"Right now I am just beginning to see these things that are very negative. They are so encrusted but I hope to start to break them down. I am very, very thankful for this opportunity and this chance. It is one in a million!"

Chapter 11

Unholy Six Testimonials

June 20, 1979

Dan: "For those of you who are new I can see there may have been some confusion on your part as to what is going on here.

"We as a group of students have all lived many negative lifetimes and we have all been in opposition to Truth and Light. Unarius is the Truth and it is the Light of the Higher Worlds and Minds who have come to this planet; not just to this one planet, but to all physical worlds at this time and we're attempting at this point in our evolution to come to grips with these negative expressions that we created within ourselves; to face them.

"With the understanding of the Science, we can accept the Principles of energy that have been given us by these Higher Minds, through the Channelship of Ruth Norman or the Archangel Uriel. We are beginning to unmask ourselves and to become aware of this other self, this higher self that wants to progress and wants to live in these Lighted Worlds. We will if we, ourselves, do not defeat that purpose and give in to that lower ego self. At this time, the ego self is very strong - particularly in this cycle. I speak from my own awareness as I have experienced it. All of us, I am sure, are going through difficult times, and if you haven't been studying these principles, then you haven't started upon your progressive evolution because it is these principles that are going to be the saving grace at this particular time in our cycle called the Unholy Six Cycle.

"Just recently, I have gone through a very negative reliving and workout and the only thing that kept me from going totally down was the understanding that there are these Infinite Minds who are working with us, channeling the Love and these energies of Intelligence to help us to come to grips with this past and to realize there is no one condemning, no one pointing fingers; that they are all working for our eventual freedom. But we have to participate in it. We have to be willing to look at that old self in an objective and truthful manner and not try to hide any facts of what we are and what we have created. So during this particular part of our session we try to give it over to testimonials. So if there is anyone who has had any testimonials of any workouts that they would like to share with the rest of us, this is the time."

Leila: "Since we started these planet filmings, I have been around students more than usual and of course I have had reactions. One particular one was that I was reacting to psychological tactics that were being used; like when my name would be said in a certain tone, I would just want to be climbing the walls. I would say something to my husband about his behavior and he would say, 'Oh, that's your imagination,' and I would become very emotional. So I finally realized that this is the way we treated Dalos - projecting psychological tricks at him and so forth. It would just cause him great distress.

"Also, during the Peacock Procession, I stood near Uriel when she was on the Isis throne and I could see her beautiful facial expressions and the way she was speaking was just so superb! I had never seen her like that before. Then when Tryonus walked up, I felt as though I was in Tryonus' clothes. This was quite a shock to me and I really had needed it because I had wanted to see myself so I could get rid of some of the blocks that I do have and I really got a taste of it that night!

"Then when we had the filming of Planet Valneza, I felt like I was in the lower astral when I saw those people made up so negatively. I know where I have been in my evolution most of the time. I saw contemptuous arrogance, I saw merciless coldness, cunning indifference, self-criticism and I came in close touch with insanity."

Kathryn: "I had the thoughts during the filming many times that we had Dalos in the force field completely under our control, so we thought we were ruling the Light now. So this kept me safe and I felt very superior. I was at the Center when Uriel came in one day and we were working on the flame dress. (Cosmic Generator). She was saying she would really like this garment to be finished by Sunday. She was getting everyone more involved in it. Then I started becoming involved in this filming and reading the chapter and I took my mind completely off the flame dress and had only a week to finish it.

"So I am finding that I hadn't been listening to Uriel, our Teacher like I should be. My consciousness wasn't where it should be. I was totally melancholy just following and doing; just totally a robot and this is not a progressive way to be. So I am finding that if I listen to what Uriel has to say and take her words to heart, then I can walk on this lighted pathway more directly. Also I know that I must walk this lighted pathway on my own from my own inner self and strive for the little hookup I can attune with. This is the way I must make it: really on my own and let the other students make it on their own and walk this lighted pathway. And we can help each other through our testimonials and being around each other and becoming involved in the different things there are here to become involved in. Thank you."

Dennis: "The last couple of days have been very helpful for me. We had been doing the filming Sunday and I realize now how this was set up for a workout for me to try to play the part of Macinus - a female role and, as I said, I looked like a very dumpy female polarity. What really rang in that Sunday, was being out in a public place being dressed up as a woman, feeling totally fake; everything that came out of my mouth felt heavy. It was just a totally negative situation. If people came by, I could not even look them in the face because I felt they would know who I really was and the feelings I went through initially, impersonating Dalos.

"So as the cycle came in, this was the experience. There was a little insecurity feeling, 'How can I pullthis off?' I was just somewhere in the ranks on the Orion Empire political scene and was pulled out for some acting ability and was given plastic surgery and given training on how to impersonate this Being. Then I was taken to the planet Idonus to set it up as a pleasure planet, to be the leader of this planet - a puppet really. We filmed much of this Monday, so in that filming I had the realization of the tremendous ego I have, the power-mania because as I landed on the planet, there was a feeling of mastership and domination of this planet. I had a device implanted that fed me from the computer and also gave me a psychokinetical ability to mesmerize people and I did this to Kathy. Kathy ruled with me for that five hundred years and I obsessed her.

"I also obsessed Steve Sandlin. He was 2nd in command and I strung him along and would project the feeling that he was going to have some additional power and then I would deflate him and I manipulated other people like this.

"I have been a puppet to all of these energies. This has put a block around my consciousness that has not allowed me to really incept much of the Light frequencies that are available if I am aware of them. There is a very confusing level of consciousness that I have because of this negative feedback.

"There is also some resentment for Stephen because he helped make me the fake Dalos and then he was also one of the people who incarnated in this experiment later and we had a confrontation in the filming that was very real, where I tried to dominate him with this psychokinetical power and bring him to his knees as I was able to do with Sandlin, as I had control of him in this negative way. But Stephen had just as strong an opposing force and I have resented him ever since and I probably resented him even before that. I am sure that wasn't the very beginning of it but it was a really crucial moment of my resentment towards Stephen and my jealousy in that he is playing a superior role.

"I am also realizing how I am in a way responsible for many of the pleasure seeking escapisms that are dominant in this world. This five hundred years of building pleasure machines and building this planet into a pleasure gem for the Empire has given a lot of force to many obsessive entities that are in full sway in this Earth world in the present. If I don't work those out, I continue to add my own negation to this force, for it is very important for me to recognize this and there is no other way than the way we are doing these films. I am very appreciative of this time here."

Leila: "Yesterday when I was using the vibrator on my neck, I had a severe pain that shot up into my head and I realized, that due to the vibronics and sounds that we used on Dalos, that this was the reliving on that.

"Also, one time Steve Anderson gave me a reading that I had controlled the elements and I couldn't relate to that at all until the night I saw the spacecraft movie where they were manipulating elements and I then realized that I had been there and had actually done this.

"I also got a very good example of my schizophrenia the other day. Uriel asked me how I was and I replied, 'Well, I feel better; I dropped a lot of garbage on the weekend.' When I said that another entity or personality came in and Uriel had to turn her head when it did so and I realized what had happened. Thank you."

Bill: "I had a note given to me by Uriel just the other day that helped me a lot. I went through the insanities the day after. I did something that I haven't been able to do before and that is, I was able to say, 'I am negative. She wouldn't say anything unless it was something that was going to help me because she loves us!'

"So I began with that although I felt it was going to take awhile to break through because she had hit a tender point in my psychic.

"The other day I said, 'Uriel, you owe me twenty-one dollars for the photographs I took.' I had given her some snapshots. She laughed and said, 'Sure, I will pay you tomorrow.' I felt great and I felt super because I would then be able to afford to film more and she was helping me in this way. The next day when I received a check from her I also found a note saying, 'You say I owe you twenty-one dollars,' and she went on and on so that it really hit home. I read it about ten or 15 times. When I read it tonight in the Center, the energies were a little different and I realized she had underlined what I had said, 'You owe me!' The whole thing was that I was photographing to obtain releasement from these karmic situations, as I had filmed her in negative situations in the past giving Uriel appearance in photographs as a historical event; the most historical event that has ever happened on this Earth which I was now attempting to present with a positive glow in the photographs to show what Uriel and the students are now doing. I really didn't realize that I was doing it for my own benefit. You know, 'Look at me!'

"She, in her all-knowing nature, hit the button to at least let the wound start to release the sickness that has built up. And it is something I went insane with yesterday, I was practicing acting for the play that we were going to film where I tell the Infinite that I don't care what it has to say; I am going to do what I want to do and that is why I am existing because I am going to do what I want to and it is in my mind and I went insane. I called it practicing but it just beat through and that was proof to me that this is what my attitude was like, 'I am going to film Uriel in this life of Dalos; I am going to make her look bad; I am going to get that position and have that place on the pleasure planet!' I really wanted a palace, I wanted my name everywhere and I wanted to be recognized and looked up to as just a great person, as great as any of the leaders.

"I even had the undercurrent starting to move where even to these leaders I was going to gang up with other people to get rid of them because, 'I just wanted to get rid of them because they are making the decisions and we are doing all the work.' I wouldn't have admitted that before - that I would have been glad to do away with some of the leaders!"

Calvin: "I have a realization about a certain aspect of the personality I present and the thing that bothers me the most about myself right now is that I can't stand having to back down. In other words, if I am in an argumentative situation, I will argue to the end instead of sitting back and saying, 'Oh, well, maybe I'm wrong.' And this is something I know I have to work with and let go of and it came to me that I had been programmed to feel this way, 'I'm not going to back down, no matter what!' I believe I also helped to program this into others during these times of the Unholy Six.

'In other words, my whole being was set up to feel that the only way I could get into a position of power was by reinstating myself over and over again and by never backing down no matter what the situation or what anybody said, etc. So I think this will help me now to realize that it is just an obsession that I have built up and when this comes in I can look at it with the attitude, 'I am the loser if I continue to reinstate this big fat ego or power play.'

"So it is an important realization for me and I know that if I keep this in mind, it is going to mean a tremendous amount more to me in the future."

Christine: "I have got to force myself to get up here. I have to say that right now, I really feel terrible. Parades always make me sick, depressed and just feeling like I want to run away. We were watching the parade in the Center today that was being filmed. I think I have tried to block the whole thing out of my mind; I am not really sure what it's for. I think it's for Uriel and I think it is for prior to Uriel announcing to the world that she is Ambassador of the Confederation when she wears her Cosmic Flame dress. I think that is what it is.

"But like I say, I am totally hazy on the whole thing because I haven't wanted to see it because to me it represents what happened during the Unholy Six and how we infiltrated other planets with our propaganda and said how great the Orion Empire was.

"And now we are doing it in reverse. We are infiltrating this earth world with the Light but it still tunes me to my past and I get very, very negative with it. So that is something I have to work on to recognize.

"Another thing that I went through just the other day was in realizing my varityper tunes me in to working on the computer planet in a strong way. I am beginning to actually grasp or understand what I had to do with that computer planet. I think I had something to do with controlling and keeping functioning a large portion of it and keeping the information going to the many planets. So, in my psychic anatomy, this computer I was working on, which the varityper tunes me in to, was so vast and it was so powerful, and I was such a part of it that I was like a robot to it because when the varityper malfunctions, I malfunction! I just feel like I am going to explode and I have really had a hard time keeping myself sane. It is like when your life falls apart. That is the only thing I can relate it as being: Your whole life is just fading from view right before you. That is how difficult it is for me when that machine starts to act up on me. Many times it has to do with my negative state of mind; it will just start working badly and it is because I am negative.

"Yesterday the carriage of the varityper was as far to the right as it could go and I pushed the return button. At the same time I knew I shouldn't return it because the carriage was already so far right that it would go left with such a degree of force that it would jar or would jam it; there was too much force when it was travelling back the other way and that is what happened. It knocked something out of kilter and from that time on the return mechanism wouldn't work, plus the spacebar wasn't working.

"Well, I just simply refused to have this happen and I kept saying, 'This is not going to break; it is not going to break!' At the same time, I felt I was going insane and then I said, 'This is a terrible reaction here. What is going on? What is this?' I saw how in the Unholy Six Cycle, having to do with the computer planet, I had so much information going out to these planets that if the computer malfunctioned, a whole day would be lost in getting information to a certain number of planets and that would set me a whole day behind and that would foul up the works on the computer planet and things would start backing up.

"I have been going through this pressure with the Dalos book. I can't seem to keep up with it and I have been trying to figure out ways to keep up with my work. It has been a great pressure with me and it's all from that past.

"Just by my refusing that it would break, I knew I was negative and I knew it could be corrected if I worked out the cycle somewhat. That helped a great deal and the return mechanism started to work properly and the spacebar started to function half way but it wasn't doing the job completely right. Then Uriel walked in and I told her I was having trouble with the spacebar. So she put her hand on the machine; she put her hand on my forehead and walked out of the room. I then started to type and everything was just fine! It was amazing! She fixed it right there with her psychokinetical energies. So I know she can do anything.

"Anyway, I realize the reason I am so blocked off is because working on that computer planet as I did, I have just hooked myself up with machines down through my evolution ever since then and I haven't been able to function without one. And when the machine breaks, I break and that is a terrible state to be in, when you cannot stand on your own two feet without some type of apparatus to make you work - to make me work.

"So one other realization I had was that the reason I have a hard time tuning in to the Lens, which is the abstract Intelligence of the Universe, which is actually the Higher Minds, a globe of positive energies that we can symbolically tune in to - just the idea of the Lens . . . I have the idea of a planet in my mind and it tunes me in to the computer planet which is the negative force's answer to the positive counterpart of the Infinite; that is, the negative counterpart of the Infinite. That is what blocked off infinite numbers of souls and I was working on it, so therefore I am blocked off.

"I just hope, with my constant recognition, that in this lifetime I will be able to work out some of that negation."

Keymas: "I had a dream this morning that I want to use for an ego inflation but I know it wasn't meant for that. I was in the Center by the New Age City, when all of a sudden we were filming the students who were all gathered around Tom and he had me sit down and said, 'Alright, just let the Brother come through and speak His words, just relax.' And I said, 'My God! I am not ready for this!' Right there is part of it - not being ready and not being prepared. I sat down and was very nervous. I said, 'I can't channel, or can I?'

"I began to think, 'Well, maybe I can if I relax. What is channeling about but just speaking another's words or speaking another Being's words and being able to get yourself out of the way?'

"So I became relaxed and I quested, 'Who will the Brother be?' and all of a sudden it came in, 'Kuthumi.' And I got the chills all over just like when Uriel walks into the room. So I knew this was true and it was very familiar.

"In another dream state I had an association just recently with Kuthumi or with that name because it was so familiar and I have never had a rapport before with Kuthumi. So I began to speak but nobody was listening; everybody was talking and it was kind of upsetting. I began to speak and the Brother really wanted to speak through and I wanted to speak what I felt the Brother wanted to speak. And this is my problem when I talk to somebody about the Light and about the Truth; I tell them what I think they should know and I feel the Brothers there because I get really hot. But I feel this backing up of energies, these energies that the Brothers are trying to bring through but I don't let them through because I want to dispense what I want to dispense. I will tell a person about this because I think this is conducive to his life, which isn't right and very few people listen to me. Then when they do, I become very egotistical; I get this tremendous welling up and feeling that it is 'my' wisdom - the wisdom that I have attained when it isn't really.

"As I was writing this down it came to me really quickly. I realized I had that dream and began to write it down and there was this incredible force that wanted to use it as an ego inflation. I wanted to say, 'Well, this is my future and I must really be prepared,' and it was so strong. But I don't know how it happened; there is this other entity, this other force and you can't really separate either of them. They came through and this made me write truth; this made me write down what the truth was. I felt, 'I can't write this!' and as I was writing it I was arguing with myself, saying, 'I can't write this, this is going to pollute the whole thing; it is going to tear down what I want to make this out of.' But it came out and I am glad now, because in the future maybe I will be able to. I know I will in the future; it is my future to be a channel in a more direct way but not until I realize it is not I. And it was very helpful to me and it was very real and vivid. Thank you."

Note

January 20, 1980

Uriel: "As has been the custom in these class sessions for the group of students who have gathered together in the San Diego area, that there is heard either a dissertation by the Brothers on the Inner or that there is played from the vast collection of tapes, one of the former lectures. Then during the last half of the meeting, the students are given an opportunity to voice on tape their testimonials or relate regarding the need for healings, etc.

"To say the least, these sessions have been most beneficial in all respects. At this point in time, each and every student has and is making progressive headway in his evolution. This means his total direction has been changed so that he has now discontinued his reverse or downward direction and is now adding to his more positive, forward, progressiveness and which, in itself, is no small achievement. This could be accomplished only due to the constant and never-ending inner help by the Brothers on the Inner, whose prime purpose is to aid with their power and healing energies. These, of course, are the selfless Ones who serve in this devoted way, all mankind.

"The home study student should realize that he too can experience these same benefits. He can overcome his past and move forward into his future, freeing himself of these fetters and bonds of the past. And it may be well to mention that of late there have been students who have unlocked certain past negative ties that have held them thusly bound in a negative way, opposing the Light for as long a time as over a million years! It is quite difficult to imagine that some deed expressed or some decision made mentally, could actually waylay one from a progressive way for lo, this great, vast length of time; yet this is true. There is no time limit wherein these negative energies are exhausted or wear out their potency or strength; in fact, they are only strengthened by frequent physical incarnations in this negative living and reliving.

"So when one can experience such a freeing as those who have recently come into this awareness of instances where they realize that they have turned their backs on the Light these many thousands of years ago and are now reversing this deed and direction, it is indeed a life-changing and wondrous thing. The person suddenly takes on a new attitude; instead of being hostile, resentful and putting out a negative energy, stepping up and defending his ego at every turn, he now actually, with such freeing and healing, is mentally working for the Light!

"I am most happy to say and to see some of these students who have actually fought me (the Light) psychically and mentally so desperately and for so long a time actually to change so definitely, it is a joy to behold and we know those souls shall henceforth move forward in a progressive way in their evolution. They are now Light Bearers! Need we say more?

"I am simply desirous of having you souls who study at home to know that it can happen to you - and it will. You simply need to be persistent and positive and know that your 'great day' of freedom may be just around the corner. There are no incurable problems! Incurable people, yes! But Principle is valid and does work when applied. So keep the faith and never say, 'Now wasted am I,' or 'I am incurable.' So long as you study and strive to apply Principle, We Brothers are ever with you to help in the ways of power and enlightenment; to project into your consciousness the pictures of your past when you are so ready to observe them.

"Our help, Light and Infinite Love are ever forthcoming to all so attuned to this Word. And you, too, shall know the true meaning of what it is to drink from the Pure Waters of Spirit and to live in the Consciousness of the Infinite."

OTHER UNARIUS PUBLICATIONS

Pulse of Creation Series	
The Voice of Venus	Vol. 1
The Voice of Eros	Vol. 2
The Voice of Hermes	Vol. 3
The Voice of Orion	Vol. 4
The Voice of Muse	Vol. 5
Infinite Concept of Cosmic Creation	
Infinite Perspectus	
Tempus Procedium	
Tempus Invictus	
Tempus Interludium	
Bridge to Heaven	
Infinite Contact	
Truth About Mars (1955)	
The Elysium	
The Anthenium	
*The True Life of Jesus of Nazareth	
*Little Red Box (Sequel)	
Cosmic Continuum	
Who is the Mona Lisa?	
Conclave of Light Beings	
Tesla Speaks Series	
Scientists	Vol. 1
Scientists & Philosophers	Vol. 2
Scientists & Presidents	Vol. 3
Other World Contacts (3parts)	Vol. 4
Other World Contacts (Cont'd.)	Vol. 5
Crystal Mountains, Cities & Temples	Vol. 6
Countdown to Space Fleet Landing!	Vol. 7
The Masters Speak (2 Parts)	Vol. 8
The Masters Speak (Cont'd.)	Vol. 9
Whispers of Love	Vol. 10
Keys to the Universe	Vol. 11
Martian Underground Cities Discovered	Vol. 12
The Epic	Vol. 13
By Their Fruits	2 Vols.
Recorded Tapes of Above Teachings	
Unarius Light Magazine	
Lemuria Rising	4 Vols.
Decline and Destruction of the Orion Empire	Vol. 1
Have You Lived On Other Worlds Before?	Vol. 1